Information Systems Engineering and Management 68

The book series "Information Systems Engineering and Management" (ISEM) publishes innovative and original works in the various areas of planning, development, implementation, and management of information systems and technologies by enterprises, citizens, and society for the improvement of the socio-economic environment.

The series is multidisciplinary, focusing on technological, organizational, and social domains of information systems engineering and management. Manuscripts published in this book series focus on relevant problems and research in the planning, analysis, design, implementation, exploration, and management of all types of information systems and technologies. The series contains monographs, lecture notes, edited volumes, pedagogical and technical books as well as proceedings volumes.

Some topics/keywords to be considered in the ISEM book series are, but not limited to: Information Systems Planning; Information Systems Development; Exploration of Information Systems; Management of Information Systems; Blockchain Technology; Cloud Computing; Artificial Intelligence (AI) and Machine Learning; Big Data Analytics; Multimedia Systems; Computer Networks, Mobility and Pervasive Systems; IT Security, Ethics and Privacy; Cybersecurity; Digital Platforms and Services; Requirements Engineering; Software Engineering; Process and Knowledge Engineering; Security and Privacy Engineering, Autonomous Robotics; Human-Computer Interaction; Marketing and Information; Tourism and Information; Finance and Value; Decisions and Risk; Innovation and Projects; Strategy and People.

Indexed by Google Scholar. All books published in the series are submitted for consideration in the Web of Science.

For book or proceedings proposals please contact Alvaro Rocha (amrrocha@gmail.com).

Anthi Katsirikou
Editor

Advanced Applications on Qualitative and Quantitative Methods in Information Science

Selected Papers from the 15th QQML International Conference

Springer

Editor
Anthi Katsirikou
University of Piraeus Library
Piraeus, Greece

ISSN 3004-958X ISSN 3004-9598 (electronic)
Information Systems Engineering and Management
ISBN 978-3-032-01263-0 ISBN 978-3-032-01264-7 (eBook)
https://doi.org/10.1007/978-3-032-01264-7

This Springer imprint is published by the registered company Springer Nature Switzerland AG
The registered company address is: Gewerbestrasse 11, 6330 Cham, Switzerland

Preface

The book examines crucial issues on the library theory and practice. It is a matter to investigate how things change and how rapidly. During the pandemic of COVID-19, academic libraries adapted their services, products, and staff. Currently, questions arise on the lessons learned and the transformation. To this end, a survey was carried out among Portuguese higher education librarians based on the *Standards for Distance Learning Library Services* (ACRL, 2016).

Furthermore, modern academic libraries seek to comprehensibly organize and present a universe of information to students and researchers. This study examines the organizational charts of a small but broad subset of libraries at accredited American institutions of higher education to survey how academic libraries divide their roles and what roles librarians serve within them. A sample of 100 institutions was selected from a population of almost 4000 to represent community colleges, liberal arts schools, Masters colleges and universities, and R1 universities from across the United States. The outcome will speak to how academic libraries are actually structured to meet students' and researchers' information literacy needs.

Higher education institutions are expected to take an active role in a more entrepreneurial climate. This paper investigates how library management develops entrepreneurial activities in terms of idea generation, opportunity recognition, opportunity development, venture launch, and impact measurement. In addition, it explores the driving factors for successfully implementing entrepreneurial initiatives and determines the challenges and solutions library administrators face in meeting the goals of entrepreneurial activities. The results will help library management and school administration understand and recognise the stimuli that can drive sustainable entrepreneurship in academic libraries.

Writing research papers for publication in international academic journals is necessary and important for researchers. Specific academic writing education programs are being developed in various academic libraries to promote academic writing. This study aims to evaluate the effect on improving the writing skills for researchers attended the short-term training course. The results show that researchers improved their writing skills through the short-term training, which show significant difference between the experimental and the control groups. These results are useful for an academic library to explore educational activities on academic writing. This study provides pioneering evidence for testing the effectiveness of the training on academic writing on promoting the writing skills for English-as-a-second-language (ESL) researchers.

As we approach the 50th anniversary in 2024 of first Information Industry Association president Paul Zurkowski's 30-page information literacy founding report to the U.S. Commission on Libraries and Information Science, Washington, DC USA where he publicly identified, named and defined information literacy (IL, InfoLit), it is important to reflect on the roots, development and future of IL. Reflections are based on

11+ years learning, working and traveling with Mr. Zurkowski until his passing in January of 2022. Mr. Zurkowski's "Report to the Commission" "The Information Service Environment Relationships and Priorities" that launched IL in 1974 at the dawn of the Information Age as part of the National Program for Library and Information Services designated as "Related Paper Number 5", is even more relevant today and will be presented and explored including his quotes on democracy, active citizenship, and libraries in context of IL. Lastly, the key events that led to starting an international professional nonprofit educational public charity association to represent IL professionals worldwide, the growing international IL community and how to increase critical thinking through IL, the Universal Information Literacies Association (UiLA), will also be presented.

By highlighting the affinities between Public History and libraries as experts in audience engagement, the paper proposes activities and strategies, inspired by the Public History approach, that libraries may perform – also in cooperation with other GLAM communities – to promote historical consciousness and cultural heritage as a vehicle of knowledge and social inclusion, therefore enhancing the connection between libraries and democracy, so that it may become more effective.

The life cycle of information begins with an information need with a special role in libraries. Knowledge of informational sources and especially of electronic sources are recognized as vital factors at all levels of research with higher literacy needs than any other time. Through digital marketing, an organization or agency, with appropriate management, can effectively communicate with its environment and audience by utilizing digital tools and social networks, and through the balance of these and traditional communication channels. In this context, the participation of collaborative networks and ongoing information literacy programs based on International Standards shape cultural values and abilities in users to produce independent, directed scientific research. The Big Six Skills method can be a dynamic tool for the development of information literacy skills with the basic one being that of the correct use of information with basic steps being contact with information sources (hearing, sight, reading) and the correct identification of similar sources. To strengthen the role of libraries in IP education, and explore how they can connect with the needs of the economy, present research explores the awareness about IP skills and competences of non-lawyers (engineers, scientists, business managers, etc.) within labor markets and compares preliminary findings of two project partners, Croatia and Slovenia, who have comparable educational contexts.

Contact tracing and other pandemic control measures posed additional dangers to marginalized communities not faced by members of dominant communities. This is because marginalized communities are more likely to be surveilled and policed by others. While there is almost universal agreement that privacy rights are critical, there is significantly less agreement regarding what they are, whom they protect, and the appropriate remedies for privacy harm. This paper will discuss some of the significant areas of law impacting LGBTQ+ students' privacy, making clear the immense legal complexity that both libraries and librarians must navigate to protect LGBTQ+ students' privacy rights.

In the era of globalization, there is a pressing need to inform college students about intellectual property rights and related law. Intellectual property is an integral part of the sports world. Sport and its linked activities are developing along with several new phenomena, difficulties, and risks, requiring strict observation and legal compliance. While copyright and trademarks serve as the foundation for many industries (such as broadcasting, fashion, and brand value), it is important to use other types of IP rights to safeguard content and creative work. The current study was conducted to determine the level of IP rights (IPR) knowledge among college students at University of Zagreb Faculty of Kinesiology. The descriptive research method was used for this investigation: the study was conducted through a questionnaire. The outcome of this study will inform projects in the field of Intellectual and Industrial property which aim at forecasting the needs of the future labor market regarding Industrial Property while introducing education and proposing curricula in this field.

In this article, the authors point out the importance of the building that is believed to be the oldest residential building in the town of Kraljevo. This building is known as the Adzic House. This cultural monument is a good example of how the cultural heritage of past times is combined in the best way with contemporary creativity and cultural trends of the present. In 1976, conservation and restoration work were carried out, returning the house to its original appearance. Since March 2001, the employees of the publishing department of the Public Library "Stephen the First Crowned" in Kraljevo and the members of the Kraljevo Book Club have been working on the premises of the Adzic House.

The paper is aimed at determination of ethical factors which can shape value-sensitive information services of academic libraries. It is based on analyses of selected theoretical frameworks of information ethics studies and on an own empirical Delphi study of experts in the Slovak and Czech Republic. In the analyses we applied discourse analyses, content analyses and conceptual modelling. We present a resulting conceptual model of ethical factors of human information interactions. We proposed the inclusion of the identified ethical factors into models of information literacy and human information behaviour. In conclusions, we introduce a concept of value-sensitive information services of academic libraries focused on verification of information.

Over the last few decades, there has been an increased interest in flexible electronic devices due to their Due to flexible electronics is an emerging and multidisciplinary research field, the field of flexible electronics still has many challenges in terms of materials, fabrication and specific applications, before it becomes part of our daily life.

National Science Library (NSL) is a research library attached to Chinese Academy of Sciences (CAS). NSL have a subject librarian's team with diverse subject background. One important role of subject librarians is to provide library services to support scientific research in institution of CAS. The idea of change in research libraries and the development of information technologies motivates subject librarians to think about ways of library precision services to support researchers, especially in an emerging and multidisciplinary research field. The results showed that the annual number of publications in the field of flexible electronics worldwide has increased significantly during past 20-year. The top three countries in terms of the number of publications were China, USA and Korea. The top three contributing institutions included Chinese Academy of Sciences, Tsinghua University and University of California System.

Although it was a case study of an emerging research field, some useful information had been revealed. Firstly, the accuracy of datasets was critical success factors quantitative analysis. Quantitative analysis using bibliometrics and visualization was still an effective method to help researchers comprehensive understanding current research trends, especially for emerging and multidiscipline field. And the researchers were satisfied with the results of current research trends by quantitative analysis. Meanwhile by cooperating with researchers, subject librarians could accumulate best practice experiences and some useful insights in different fields. Finally, it was also important for research libraries to make more strategies precisely providing information services to satisfy researchers' needs.

This paper assessed the scientific and technological performance and impact of Advanced Photon Source (APS) in USA as an example, especially from the perspective of 68 simultaneously beamlines/ end stations, which are the key components of research infrastructure and the frontier of scientific service. The main dataset was constructed based on the academic paper records allocated from the official website, including the output using APS and by the staffs of APS focusing on the establishment, development and upgrade of APS during 1985 to 2019. Furthermore, several issues about the sustainability of large facility were discussed briefly: (1) the necessary of upgrade; (2) the challenge and alternativity from new disruptive experimental instruments and technology; (3) supporting trans-disciplines scientific research effectively; (4) international cooperation and different access demands in various countries user.

In this chapter, we discuss the primary methodological and ethical aspects related to studying conversations that take place on different social media platforms, such as Facebook, discussion boards, and dark web forums. The presented views are based on the authors' previous empirical research, which has focused, among other things, on social media discussion and reactions to politicians' posts, discussions on drug websites on the dark web, and the information practices of socially withdrawn (hikikomori) in picture board discussions. In addition to theory and methodological considerations, we briefly present examples based on our research. In summary, protecting anonymity and confidentiality are the most important things that should be remembered in the research on discussion groups, picture boards and, above all, the dark net.

The constantly changing economic and social environment, in connection with the recent epidemiological crisis, imposes the focus of libraries and cultural organizations on their sustainability. In this context, alliances are an important resource along with an integrated cultural management strategy, adapted to the ever-changing environment.

The case study focuses on collaboration between libraries of heterogeneous bodies with a common theme, which is a modern and innovative form of consortium, as the common object addressed by the participating heterogeneous members, contributes to the development and planning of joint actions. The object of this presentation is the creation of strategic planning of Maritime Libraries Network (MarLiNet), a collaborative scheme between university libraries and private libraries of cultural institutions. In the context of this process, the analysis of the internal and external environment of MarLiNet was carried out, using the PESTLE and SWOT analysis to determine the objectives of the Network for the drawing up of the strategic planning. The research process was based on the positions and statements of the members of MarLiNet and in the observance of specific international methods of strategic analysis, to render with the greatest possible clarity and accuracy the strategic planning of the case study. The objective purpose of the strategic planning is to capture as effectively as possible the long-term direction of the cooperative network and to examine and study the future progression of MarLiNet.

Anthi Katsirikou

Contents

Information Services

Academic Libraries

Where Are the Librarians in the Academic Library? Comparing Organizational Charts from Higher Education Institutions in the US

Monica Carroll(✉) and Elizabeth Marathas

University of Illinois, Urbana-Champaign, IL, USA
monicac3@illinois.edu

Abstract. Modern academic libraries seek to comprehensibly organize and present a universe of information to students and researchers. Almost two-thirds of American adults over the age of 24 have attended at least some college and had access to the information literacy instruction that an academic library provides. There is a significant variation in US higher education institutions, however. A quarter of annual college enrollment is in community colleges. Another two-quarters of students are enrolled in colleges and universities without high research activity. Nevertheless, studies of how academic libraries are organized and managed often focus on R1 or very high research activity universities. This study examines the organizational charts of a small but broad subset of libraries at accredited American institutions of higher education to survey how academic libraries divide their roles and what roles librarians serve within them. A sample of 100 institutions was selected from a population of almost 4000 to represent community colleges, liberal arts schools, Masters colleges and universities, and R1 universities from across the United States. The organizational charts were collected from library websites or personal solicitation and converted into network data. Annual federal survey responses were also collected. Text, numeric, and network analysis will be used to explore the division of library labor and the position of librarians within that structure. The outcome will speak to how academic libraries are actually structured to meet students' and researchers' information literacy needs.

Keywords: Library Organization · Data Models · Qualitative and Quantitative Methods

1 Introduction

Information literacy is critical to modern citizenship as access to publishing has exploded for both content creators and content consumers. The ability to find the information we need, evaluate its trustworthiness, and use it in our lives are essential skills. Although they are fewer than public libraries and far fewer than school libraries, academic libraries are positioned to play a major role in information literacy instruction. Students opt in to higher education and express their information needs through degree enrollment and

A. Katsirikou (Ed.): QQML 2023, 68, pp. 3–17, 2026.
https://doi.org/10.1007/978-3-032-01264-7_1

career pursuit. This intentionality is a boon for the development of focused, contextual instruction and support. And while higher education is not available to everyone, almost two-thirds of Americans over the age of 24 have attended at least some college putting them within the reach of academic libraries (National Center for Education Statistics, 2023).

The American Library Association's (ALA) Association of College and Research Libraries (ACRL) envisions contributing to a "thriving global community of learners and scholars" and as one goal of its strategic plan "Equip the academic library workforce to effectively lead, manage, and embrace change, advocate for their communities, and serve as a catalyst for transformational change in higher education" (ACRL, 2006). This has become an essential goal, as the information landscape has undergone dramatic changes over the past few decades. Academic libraries have been at the epicenter of information upheaval: media formats multiplying (Analogy, digital, social), concepts of intellectual property rights continually evolving and branching (access vs. ownership, fair use), long-standing models of authority being challenged and reconceived (Critical librarianship, Retraction Watch), and subject areas dividing and recombining. Perceived as staid and hide-bound, academic libraries have had to be responsive and lithe. These capabilities are manifested in the workforce and the organizational structure that supports the manifold functions of an academic library.

There is no one model of how best to structure an academic library and the services it provides, though. Some schools serve small, local populations of professional certification seeking students, others serve large national and international populations training for advanced degrees. The information needs of a sonogram technician in training may be significantly different from those of a professor of economics and the libraries that serve them may have responded differently to changes in information exchange. The variations in student population and training objectives of higher education institutions confounds comparisons among schools. The 50-year-old Carnegie Classification is a method of accounting for the wide differences in school missions and communities begun by the Carnegie Foundation for the Advancement of Teaching in 1973 (The Carnegie Classification of Institutions of Higher Education, n.d.). The classifications are updated periodically, currently every 3 years, with the latest in 2021. This classification is the basis for the term R1 to describe a category of university that awards a minimum number of scholarly doctoral degrees, spends a minimum amount of money on research and otherwise distinguishes itself as a "very high research" institution (Indiana University Center for Postsecondary Research, 2021). In the latest classification, 146 of almost 4000 accredited institutions of higher education were classified as R1 institutions. These universities have commonly recognized names, and their libraries have large collections, large staff, long histories and support a large amount of federally funded research. It is not surprising that studies of library organization have focused on these institutions. Nonetheless, over 15,000,000 currently enrolled higher education students (78%) are not attending an R1 institution and are presumably being served by 3500 libraries and learning resource centers (National Center for Education Statistics, 2021).

This paper presents a study of the similarities and differences in organizational structure and librarian roles in a range of classes of academic libraries. This study looks at the organizational charts and job titles of academic librarians at a small set of 2-year colleges, 4-year colleges, Masters universities and R1 universities to explore how academic libraries are distributing the work of training and supporting a thriving global community of learners.

2 Literature Review

Studies of academic library organization date back to the 1940s with Wilson and Tauber (1945) and Lyle (1961) detailing many aspects of academic library administration, including organizational charts from example institutions. As the number of institutions of higher education grew and their collections grew as well, there has been a steady interest in how academic libraries grapple with this change from an organizational point of view. In an 'ALA at 100' retrospective article on academic library organizational patterns, Dunlap (1976) points out that the academic libraries are responding to a range of forces and that community and liberal arts colleges, undergoing different forces, will need different organization. But this statement has not been followed by much research on organization outside of R1 schools, either within the paper or in others.

While studies or organization have had a restricted sample, the methodology for studying library organization has been a little more diverse. ITHAKA, a non-profit organization studying the academic community, is a prolific sponsor of academic library studies that interview and survey library directors. A 2016 report on the organization of research libraries (Schonfeld) used interviews with the heads of 18 research libraries to explore perspectives and predictions of structural change, all but 1 representing R1 institutions. But ITHAKA S + R U.S. Library Surveys, based on nationwide surveys library directors, cover a much larger sample size and include Baccalaureate, Masters and all classifications of Doctoral institutions. The latest ITHAKA S + R U.S. Library Survey (Hulbert, 2022) asked questions on a range of organizational issues such as communicating the value proposition of libraries to university senior leaders and what are seen as high priority goals for the library and what areas they envision growth in. These can be insightful takes on the organizational direction of academic libraries but come exclusively from the lead administrator. In addition, the latest report mentions that most of its respondents had been in the position for 5 or fewer years. So, while the institutional sample is wide, the organizational vantage point is very narrow.

Case studies are another common method of discussing academic library organization, often in the context of significant change and some of these studies have looked at libraries outside of very high research institutions. Novak and Day (2015) reviewed several published studies of academic library reorganizations to contextualize the organizational change happening at their own R1 university. Yoose and Knight (2016) describe the transformation of the organizational structure of a baccalaureate college library from a traditional model (Technical Services & Public Services) to a cluster, or modified team model and how this transition has been perceived by the staff. These case studies delve deeply into the faults of one organizational model and the benefits of another. But one conclusion of the Yoose and Knight article is that changing an organizational structure does not automatically change the organization's culture. Case studies provide insight into the organizational considerations, the plan and the implementation. It is less clear what reality settles in to.

Finally, a method in use since at least Wilson and Tauber in 1945, organizational charts can be collected and compared. These studies are not common, but Peper (2022) analyzed the library organizational charts of 61 members of the American Association of Universities, a membership organization of very high research universities. The focus was on senior leadership positions and the positioning of scholarly communications departments within the structure. Peper also points out that the Association or Research Libraries periodically collected organizational charts from their members for their SPEC Kits from 1973 to 1991 to help in the discussion of how libraries were organizing themselves to respond to change. These charts show titles and groupings of areas of librarianship. They may include all departments of the library and actual formal relationships. A drawback of this data source is that it will not show informal relationships and, due to their structure, it imposes a hierarchy that may not exist in practice.

Organizational charts are also difficult to use in the graphical form they come in. The authors worked to develop their own method of data transformation on these charts, unable to find published methods. This study looks at librarians and their job role within the library. Therefore, one step in the model is to recognize which positions are librarians. A second step is to develop standard terminology for the areas of specialty for librarians. And a third step extract data that places this position within the organizational structure.

The resulting dataset will answer how modern American academic libraries are deploying librarians to achieve their missions and improve the information literacy of our communities.

3 Methods

3.1 Sampling

Libraries were selected randomly from four groupings of Carnegie Classifications; 25 schools from each of four groups: community colleges from (Basic2021: 1–3); four-year colleges (21 and 22); master's universities (18–20), and very high research universities (15) for a total of 100 schools. We further constrained each sub-sample to include equal numbers of schools from each of 5 regions: Northeast (IPEDS region 1 and 2); Southeast (region 5); Midwest and Plains (region 3 and 4); Rocky Mountains and Southwest (region 6 and 7), and Pacific (region 8). Both public and private non-profit institutions were included in the sample (Fig. 1). Appendix 1 lists all the schools that were selected.

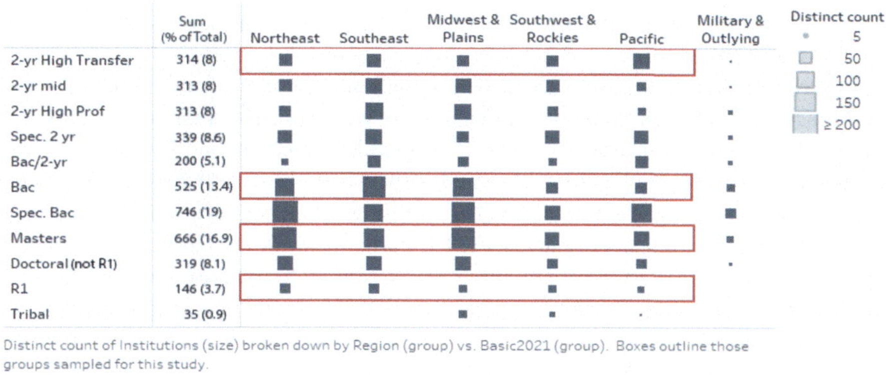

Fig. 1. Distribution of U.S. institutions of higher education by region and Carnegie Classification

3.2 Data Collection

Data were collected from three sources. The U.S. Department of Education releases annual survey responses on its Integrated Postsecondary Education Data System (IPEDS). Responses are mandatory for participation in any Federal financial assistance program. The most recent data was collected for the 2020–2021 academic year including enrollment, full time equivalent enrollment, and total degrees awarded. The human resources component of the survey specifically asks for the number of librarians, library technicians and archivists/curators, based on Bureau of Labor Statistics (BLS) definitions, employed at the college and whether they are part-time. In addition, library organizational charts from institutional websites were collected in late fall of 2022, where available, and requested from institutions where they were not available. Some academic libraries communicated that they were too small to require a chart but could communicate the organizational structure in words. In all, we attained organizational information on 54 of 100 institutions. For the remaining institutions we collected directory information from the institutional website except in two cases where even directory information was not publicly available.

3.3 Organizational Chart Data Transformation

The qualitative data collected consisted of Job titles from organizational charts and directories. We divided job titles into two fields: Job Title (e.g. Associate Professor, Library Technician); and Descriptive Area (e.g., Access Services, Special Collections). To simplify comparison, we proceeded with an iterative approach of splitting compound titles and areas such as Librarian Professor, ignoring some descriptive modifiers (Senior, I, II, Library) and combining some terms (books and monographs, or research services and research support, or reader services and public services). We then designated job titles as either Librarian or non-Librarian, such as administration, technicians and non-librarian professionals (i.e., programmers). We used BLS (2018) and College and University Professional Association for Human Resources (CUPA-HR, 2020) descriptions to help guide classification. The one major change from BLS categories is including archivists and curators with librarians. Head administrators of academic libraries are commonly degreed librarians. Where it is relatively clear that the role is primarily administrative, i.e., University Librarian, the position was marked as non-Librarian.

Descriptive area fields were cleaned to reduce the number of unique fields (over 1500 for about 3600 positions). Compound areas were split into individual fields, for example Research and Instruction became two fields: Research, and Instruction. Near synonyms were resolved; Teaching became Instruction; Public Services became Reader Services; Digital Pedagogy became Online Learning. Three sources were used to help guide our reduction of specialty areas. ACRL partnered with the College and University Professional Association for Human Resources in 2020 to revise the classification and description of specific positions within the BLS Librarian classification (American Library Association, 2008). Ithaka S & R's 2022 survey of academic library directors, split library staffing into 20 categories (Hulbert, 2022). Finally, another model considered was from a paper by Koizumi and Widdersheim (2019) that looked at patterns in specializations in academic libraries. Their model was, itself, based on a paper by Cox and Corrall (2013) looking at new specializations in academic libraries. Consolidation continued until the number of area fields was reduced to 300.

Organizational chart data was tabulated as directed networks with each box as a node and assigned its own Box ID and the lines of reporting as edges. Boxes that described units and not individuals were noted. In addition, because of the hierarchical nature of organizational charts, the hierarchy level of each box was noted with the primary administrator of the library as level one with numbers increasing with distance from the primary administrator.

4 Results

Organizational charts were found for or provided by 54 of the 100 schools chosen (Table 1). For many community colleges with small staffs a textual explanation was deemed sufficient to embody a chart. Of the remaining 46 schools, 43 had library staffing information including job titles on their websites or responded in email with job titles and for the remaining three schools, North Greenville University library does not share staffing information; Yellowstone Christian College is quite small and undergoing staffing changes, and Harvey Mudd College shares a library with 6 other area colleges so while staffing information is available, it was not equivalent to the data from the other schools. Another unanticipated challenge was that several schools within our random sample are online schools (in parentheses, Table 1). These schools do not maintain their own physical library, nor their own independent library staff. Four of the five schools direct library inquiries to an affiliated in-person school. One, University of the People, relies on subscription packages and volunteers. These schools were counted as directory information available with zero staff. Organizational charts do not always include the full staff, especially in larger institutions. Directory information was collected to supplement organizational chart data in 14 cases.

Table 1. Summary of institutional information collected

	2-yr	Bac	Masters	R1
Just Organizational charts	13	7	10	10
Just Website directory	12	14 (2)	14 (3)	3
Org chart + directory		2		12
No data		2	1	

Data in parentheses are for schools with no library

These schools yielded 3306 library employed individuals or vacant positions. The analysis began with over 600 different job titles and 1600 different areas of focus. The number of job titles was consolidated to 140 choices and the subject areas to 300 choices with individuals able to have different combinations of these terms. The original titles and subject areas as well as the filtered values will be available in an open data set.

Categorizing some positions as librarians was easy for some job titles such as Librarian, Professor, Cataloger and Archivist (for the purposes of this study Curators and Archivists were included with librarians). We also included Chair, Director and Head. Conversely, Technicians, Specialists and Assistants were tabulated as Library Support Staff as were Managers and Supervisors. University Librarians, College Librarians and Deans were counted as Administrators. Business Office, Human Resources and Communications were also included with Administration. And Information Technology positions such as Programmer and Systems Administrator, that were not also listed as librarians were excluded from the librarian group as outside professionals. Some titles, such as Lead and Coordinator defied easy generalization and were excluded from the librarian group. A summary of the resultant counts of librarians is included in Table 2 grouped by Carnegie Classification. The sum of student enrollment at the subject institutions (IPEDS, 2021), as well as the sum of the total number of degrees awarded in 2020 is included as well. These numbers were used to calculate an average of the number of librarians per enrolled student and per degree awarded.

Table 2. Summary of librarian data collected by institution

	# of academic libraries	Detected librarians	# that are part-time	Cumulative degrees awarded	Cumulative fall 2020 enrollment	Avg, degree/librarian	Avg students/librarian
2-yr Hi.	25	156	56	30,405	181,952	194	1273
Bac	21	66	0	8267	37,458	125	605
Masters	21	123	1	35,888	141,970	291	1166
R1	25	1025	1	175,183	635,470	170	619

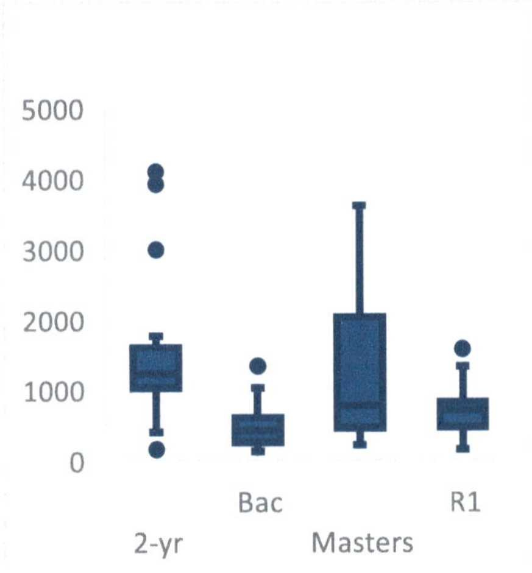

Fig. 2. Distribution of enrolled students per librarian at each institution grouped by class

While these groupings do allow an interpretation of general trends, they do disguise a tremendous amount of variation among schools within our sample. The distributions of librarians per enrolled student within each class are displayed in Fig. 2. The difference between these distributions of librarians per enrolled student is statistically significant (ANOVA, $p = 0.05$).

Subsequent analysis focused on the librarian positions within the libraries. In Fig. 3 we show the top ten subject areas for librarians by institutional class along with the relative number of librarians within that class with that subject area. Where there was a tie for 10th place, all tied areas were included. We maintained both a 'no data' field where only the title Librarian was included in the data, as well as 'Library' where the area of service was indicated as the library in general. Almost half of the librarians at community colleges have no listed specialty though it is unclear if this is because they are generalists or because specialization was not included in public facing documents.

This figure also demonstrates some limitations of our mixed methods approach. For example, Instruction and Learning both occur in the top ten of 2-year schools and R1 schools. Preserving the subtext of teaching versus learning leads to lower numbers of librarians in two separate fields where they may have the same practical specialty. Another detail to point out is the category 'Branch Library'. This term includes specialized roles that referred to a specific physical institution within the larger library. It includes both subject libraries, as well as what have been referred to as hubs. Because individuals can have multiple specializations a librarian could be at a Branch Library and be a Subject Liaison, or work at a Special Collection.

Librarians could have more than one specialization. The number of librarians with more than one specialty is presented in Table 3. Librarians at 2-yr colleges have the fewest specialties. This is highly influenced by the number of librarians with no data for their area of responsibility. Otherwise, R1 institutions seem to stand out for having the largest percentage of librarians with more than two areas they are described by. The ten most common fields to be listed in the same job title are included in Table 4. Clustering begins to appear as certain areas appear multiple times: Instruction (5), Subject Liaison (4), Reference (2), and Research (2).

Table 3. Percentage of librarians with 1, 2 or more areas of specialization

# of areas	1	2	More
2-yr	80	17	3
Bac	55	38	8
Masters	59	36	5
R1	53	33	14

Table 4. Most common co-occurring areas of specialization with number of times they cooccur

Cataloging	Metadata services	35
Instruction	Learning	31
Archives	Special collection	29
Instruction	Reference	28
Research	Instruction	22
Branch library	Subject liaison	18
Instruction	Outreach	16
Subject liaison	Research	16
Reference	Subject liaison	14
Instruction	Subject liaison	12

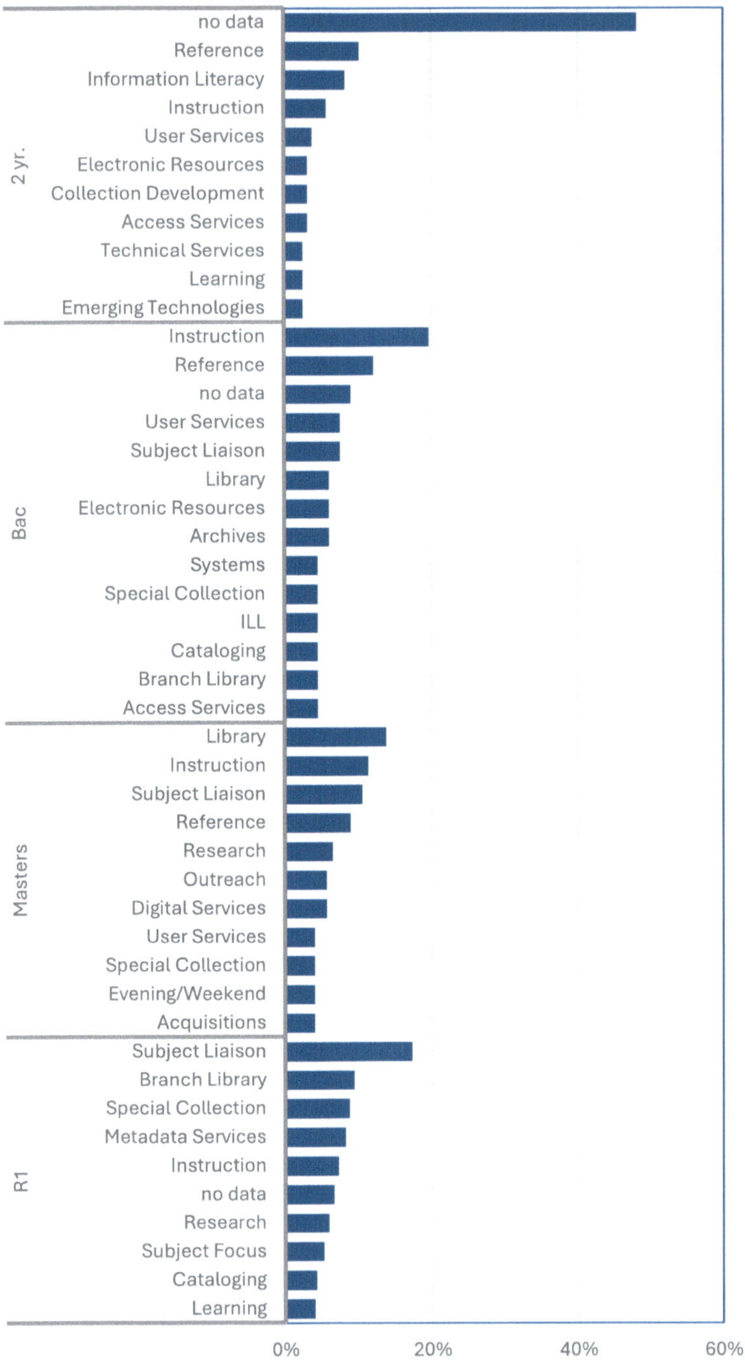

Fig. 3. Top specialty areas for librarians by class (% of total librarians in class)

5 Discussion

5.1 Online Education

This study was designed to gather a broad view of trends academic librarianship by looking at library organizational charts from a regionally diverse set of randomly selected 2-year, 4-year, Masters level and R1 universities. One of the most striking trends, though, emerged in data collection. A full 5% of our sample were online-only institutions with no librarians. The implications of this are moderated by the fact that 4 of the 5 online institutions are affiliated with brick-and-mortar institutions with physical libraries and librarians. There are different models for how higher education institutions can expand into online degrees. These four have chosen to set up distinct online degree granting institutions in federal reporting documents. Some schools keep online degrees within the colleges and departments serving campus students. Librarians, in both cases, are serving online students in addition to campus students.

This raises a question, though, of the priorities at play in moving online. Maintaining a library collection and staff is expensive. This is also true for myriad other on-campus services. There is a strong financial argument for growing online education. Yet moving services online does not reduce the need for contextualized information literacy. In fact, it probably reduces the effectiveness of contact hours, as we eliminate the magnifying effect of social reinforcement, attention focus and physical interaction. Despite this, institutions are treating online students as if they are far easier to serve, and any needs can simply be added to the responsibilities of librarians already present.

Online education has many benefits. It lowers barriers to access such as transportation, travel time, scheduling. It can be cheaper. It can be more inclusive as some differences among students and between students and instructors are not as obvious. The challenges presented by remote students, though, should not be ignored. Making online students aware of available information resources as well as training them in how to choose and evaluate sources of information are issues that important needs that can translate to informed citizenry. Online students would benefit from librarians focused on their needs.

5.2 Subject Liaisons

There have been countless debates on the most effective and efficient ways to organize librarians based on their roles–whether it be librarians as support roles, embedded librarians, or those with hybrid roles (Hoodless & Pinfield, 2018). Subject liaisons have also come under scrutiny by those who claim a functional division of library roles makes the most sense for the future of academic libraries. Hoodless and Pinfield's (2018) study uses qualitative data to explore this question more thoroughly, surveying senior managers for their decisions on a functional organizational model versus a subject-oriented model. They find that not only do subject librarians seem to remain prominent in all models, but that the discussion is more complex due to the need for balancing both functional and subject roles for librarians across institutions.

Our data corroborates these findings and makes clear that subject liaisons are not fading from eminence across academic institutions. Of the 1400 individuals classified as librarians from our data, subject liaisons outnumber the second highest role (instruction librarians) by more than double. While this may be because of the sheer number of subjects across institutions that require a liaison, this does not change the fact that subject librarians remain the most prominent role in the data.

Not only do subject liaisons make up the majority of all librarian roles in the dataset, two thirds of these librarians are designated solely as subject liaisons. Even if not included in the job title, these librarians are likely also involved in instruction, research, and reference. It is striking that the emphasis is being placed on the subject liaison, suggesting the importance of the role by the institutions themselves and those who create such roles. Regardless of whether subject liaison roles exist to support functional roles or not, they do not seem to be disappearing from academic library organizational structures (Fig. 4).

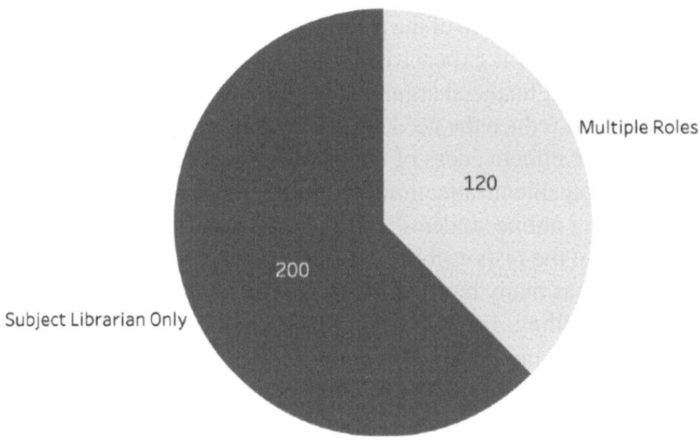

Fig. 4. Pie chart of subject librarians with multiple roles vs. sole role as subject librarian.

5.3 Digital Roles

It is clear that digital roles are becoming more prevalent as emerging technologies continue to change the academic library landscape. Cox (2016) reviews the various ways in which digital roles are defined and applied in the library setting. The nature of digital roles are described by Cox (2016, p. 133) as "experimental" and reflective of the "rapidly evolving field without clear boundaries." Our data reflect this in the large suite of digital specializations we found, including: Initiatives, Scholarship, Services, Collection, Projects, Library, Humanities, Curation, Strategies, Preservation, Research, etc. In addition are Electronic, Online and Web areas. The meanings of these terms, though, seem to be in flux. In a review of the definition for the Digital Humanities, Waters (2013) discusses several conflicting opinions on what the term means before settling on the concise definition of "the application of digital resources and methods to humanistic inquiry" (p. 2013). Taking a whole university approach we can include all digital specialties within libraries as a general category of digital resources and methods. This consolidation then raises the specialization of 'digital resources and methods' into the top ten of specializations within Masters and R1 universities or about 6% of librarians having a digital element to their job titles.

Digital scholarship is described by Sinclair (2014) as a natural extension of libraries' already existing duties within the knowledge creation and dissemination process as it becomes increasingly important to scholarly inquiry. He cites 11 academic institutions where digital scholarship hubs have become a permanent fixture in their respective library systems.

Digital librarianship is common, at least for larger universities but can have a various meanings across institutions. When looking at the cooccurrence of the word "digital" with more traditional library roles, there is no clear alignment. Digital roles were paired with 27 unique areas in the data collected, most often, with Collections and Subject Liaison roles but this could be due to the large number of these roles across Masters and R1 institutions.

Another way to look at this is to compare librarians with solely digital specializations versus those who pair a digital specialty with a traditional librarian role. What we see (Fig. 5) is an inverse of the pattern for Subject Liaisons. It is much more likely that "digital" librarianship is a modifier of role as opposed to being the sole specialty. This in turn supports Sinclair (2014) in that digital methods are just the means by which librarians perform traditional roles.

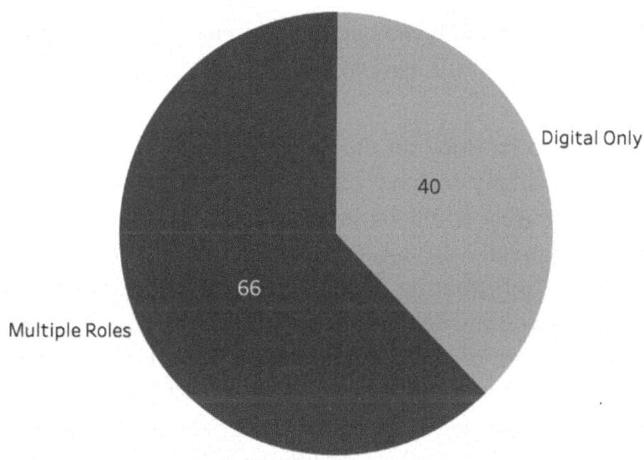

Fig. 5. Pie chart of digital librarians with multiple roles vs. sole role as digital librarian.

5.4 Organizational Level of Librarians

Librarians can be the only or most library workers at 2-yr and baccalaureate schools and the structures are, on average, fairly flat with librarians in the first two organizational levels. At R1 universities librarians are, on average, a couple steps removed from the library director (Fig. 6a and b). Another interesting pattern is the librarians included in the analysis above, have, on average, 1 direct report. One departmental librarian may supervise several librarians who do not supervise anyone. At larger institutions, any library workers lower in the hierarchy than librarians are often supervised by library support staff and not librarians.

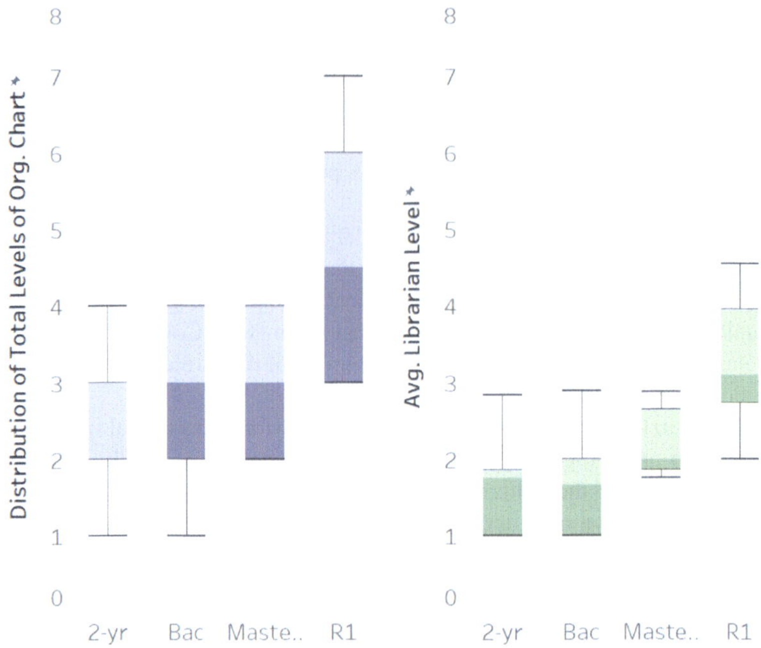

Fig. 6. Distribution of range of levels in organizational charts by class (a) and distribution of librarian levels in organizational charts by class (b)

Further work needs to be done on the data to determine whether certain librarian specializations are more likely to have reports than others. Cataloging and metadata services and Branch libraries could be areas where librarians oversee support staff, while subject liaisons typically do not supervise staff. This has implications for how academic libraries tackle information literacy. Librarians in cataloging have long been able to mobilize a work force and delegate tasks. Subject liaisons have no one to delegate to and rely on informal networks to accomplish larger tasks. Hybrid roles and team-based models would seem to be a reasonable approach to meeting the diverse roles of subject liaisons, but our data suggest that there is still a sizeable transition that needs to take place away to move from a model of single librarians meeting all the needs of specific population to a team of specialized librarians and support staff meeting the needs of a diverse population.

6 Conclusions

This study presents an initial reproducible approach for comparing organizational charts of academic libraries and applying it to a broad sample of institutions of higher education in the United States. We find that while academic libraries are embracing digital resources and methods in all aspects of the library, schools that embrace online education do not appear to see academic libraries as an integral part of the program.

Another pattern that seems to emerge is that academic libraries, especially R1 institutions, still partition the role of teaching information literacy by subject. Subject liaisons are called upon to develop their own instruction, reference, outreach and research program. Future work will explore how the organizational structure serves to bring different specializations together in teams, and how the root structure describes the larger organizational philosophy of information literacy.

References

ACRL: ACRL Plan for Excellence. Association of College & Research Libraries (ACRL) (2006, July 28). https://www.ala.org/acrl/aboutacrl/strategicplan/stratplan

American Library Association: ACRL Works with CUPA-HR to Update Academic Library Position Descriptions (2008, October 10). http://www.ala.org/acrl/proftools/personnel

Cox, A.M., Corrall, S.: Evolving academic library specialties. J. Am. Soc. Inf. Sci. Technol. **64**(8), 1526–1542 (2013). Scopus. https://doi.org/10.1002/asi.22847

Cox, J.: Communicating new library roles to enable digital scholarship: a review article. New Rev. Acad. Librariansh. **22**(2–3), 132–147 (2016). https://doi.org/10.1080/13614533.2016.1181665

Dunlap, C.R.: Organizational Patterns in Academic Libraries, 1876–1976 | Dunlap | College & Research Libraries (1976). https://doi.org/10.5860/crl_37_05_395

Hoodless, C., Pinfield, S.: Subject vs. functional: should subject librarians be replaced by functional specialists in academic libraries? J. Librariansh. Inf. Sci. (n.d.)

Hulbert, I.: US Library Survey 2022: Navigating the New Normal. Ithaka S+R (2023). https://doi.org/10.18665/sr.318642

Indiana University Center for Postsecondary Research: The Carnegie Classification of Institutions of Higher Education (2021 edition) (n.d.)

Koizumi, M., Widdersheim, M.M.: Specialties and strategies in academic libraries: a cluster analysis approach. Libr. Manag. **40**(1/2), 45–58 (2019)

Lyle, G.R.: The Administration of the College Library. Wilson, New York (1961). http://archive.org/details/administrationof0000lyle_c2f2

National Center for Education Statistics: Digest of Education Statistics (2021 edition) (2023). https://nces.ed.gov/programs/digest/

Novak, J., Day, A.: The libraries they are A-Changin': how libraries reorganize. Coll. Undergrad. Libr. **22**(3–4), 358–373 (2015). Scopus. https://doi.org/10.1080/10691316.2015.1067663

Peper, M.: A comprehensive survey of research library organizational structure. Libr. Leadersh. Manag. **36**(1) (2022). https://doi.org/10.5860/llm.v36i1.7514

Schonfeld, R.: Organizing the Work of the Research Library. Ithaka S+R (2016). https://doi.org/10.18665/sr.283717

Sinclair, B.: The University Library as Incubator for Digital Scholarship. EDUCASE Review Online (2014). https://er.educause.edu/articles/2014/6/the-university-library-as-incubator-for-digital-scholarship

The Carnegie Classification of Institutions of Higher Education: About Carnegie Classification. Carnegie Classification of Institutions of Higher Education (n.d.). https://carnegieclassifications.acenet.edu/about/

Waters, D.J.: An overview of the digital humanities. Res. Libr. Issues **284**, 3–11 (2013). https://doi.org/10.29242/rli.284.2

Wilson, L.R.: The University Library; Its Organization, Administration and Functions. University of Chicago Press (1945)

Yoose, B., Knight, C.: Clusters: a study of a non-traditional academic library organizational model. Libr. Leadersh. Manag. **30**(3), Article 3 (2016). https://doi.org/10.5860/llm.v30i3.7131

Intrapreneurship in Philippine Academic Libraries: Practices and Challenges

Marian Ramos-Eclevia[1](✉) and Carlos L. Eclevia Jr.[2]

[1] Department of Educational Leadership and Management, Br. Andrew Gonzalez FSC College of Education, De La Salle University, Manila, Philippines
marian.eclevia@dlsu.edu.ph

[2] Southville International School and Colleges, Las Piñas, Philippines

Abstract. Higher education institutions are expected to take an active role in a more entrepreneurial climate. Academic libraries are consequently becoming increasingly involved in campus entrepreneurial ecosystems. It is not uncommon for librarians who have become innovative out of necessity to invent and initiate entrepreneurial endeavours. They constantly innovate and continuously identify, analyse, and capitalise on opportunities to create a new generation of services and programs, even though academic libraries operate in an increasingly unstable environment due to rapid technological advancement and challenges from the economic, financial, and political sectors of society. Academic libraries participate in entrepreneurial initiatives for various reasons. This paper investigates how library management develops entrepreneurial activities in terms of idea generation, opportunity recognition, opportunity development, venture launch, and impact measurement. In addition, it explores the driving factors for successfully implementing entrepreneurial initiatives and determines the challenges and solutions library administrators face in meeting the goals of entrepreneurial activities. Semi-structured interviews were conducted with the directors and chief librarians of the target libraries to collect and document their entrepreneurial activities and practices. This research will contribute to the expanding body of literature on entrepreneurial librarianship by highlighting the development of entrepreneurial initiatives in academic libraries in the Philippines. The results will also help library management and school administration understand and recognise the stimuli that can drive sustainable entrepreneurship in academic libraries. It also attempts to fill a gap in the literature on developing an entrepreneurial university library in the Philippines by offering empirical data on the issues of sustainable entrepreneurship.

Keywords: entrepreneurial librarianship · entrepreneurial librarians · intrapreneurship · academic libraries

1 Introduction

Higher education institutions are expected to take an active role in a more entrepreneurial climate. Consequently, academic libraries are becoming increasingly involved in campus entrepreneurial ecosystems. As entrepreneurship in universities flourishes, academic

A. Katsirikou (Ed.): QQML 2023, 68, pp. 18–33, 2026.
https://doi.org/10.1007/978-3-032-01264-7_2

libraries have struggled to rebrand their identity as service units and learning space within the community. College students often saw the library as a quiet place to study and conduct research rather than as a resource for reference and research assistance (Lumley, 2014). Despite the technological advancements in higher education, many people perceive the library as somewhat traditional. Some school administrators fail to acknowledge the library's function, importance, and contributions to university student life and the school's internationalisation and entrepreneurship endeavours. Furthermore, Lumley (2014) claims staffing reductions, growing need to invest in or reapportion resources toward technology, and requirement to demonstrate a return of investment to students who now consider themselves paying users all impact libraries.

Given this, it is common for librarians who have become innovative to invent and initiate intrapreneurial endeavours (Scanlon & Crumpton, 2011). They are constantly innovating and continuously identifying, analysing, and capitalising on opportunities to create a new generation of services and programs, even though libraries operate in an increasingly unstable environment due to rapid technological advancement and challenges from the political, economic, and financial sectors of society (Jantz, 2012, as cited in Macdonald & Vanduinkerken, 2015). The dynamic business climate and constant technological innovation pressure libraries to continue reinventing their services to support entrepreneurship (Gupta & Rubalcaba, 2022). They further claimed that entrepreneurial libraries could serve as a nexus of "entrepreneurial activities and market knowledge" (p. 1). Entrepreneurial activity drives organisational performance (Schumpeter, 1934, as cited in Mezeh et al., 2021). Entrepreneurial activity is "the enterprising human action in pursuit of the generation of value, through the creation or expansion of economic activity, by identifying and exploiting new products, processes, or markets" (Ahmad & Seymour, 2011, p. 14).

Academic librarians may be considered intrapreneurs because of their innovative mindset and the fact that innovation is often linked with the intrapreneurship (Huang et al., 2021). Coined by Pinchot in 1978 (Franco & Pinto, 2017), intrapreneurship is defined as "employee initiative from below in the organisation to undertake something new; an innovation which is created by subordinates without being asked, expected, or perhaps even given permission by higher management to do so" (Vesper, 1984, as cited in Huang et al., 2021, p. 295). In other words, intrapreneurship is a people-centred, bottom-up approach to eliciting employee initiatives to identify and exploit opportunities and challenge established processes to create new or improve existing library services (Antoncic & Hisrich, 2001; Farrukh et al., 2019). It consists of the following main elements: "business self-renewal, new business ventures, and innovativeness related to product, service, process, or technology" (Antoncic & Antoncic, 2011, as cited in Huang et al., 2021, p. 2). Academic libraries, as service units of academic institutions, have adopted a business model that views library services as a non-profit knowledge-based enterprise and provides user-friendly information products and services (such as an online catalogue, live chat with a librarian, and reference services) for free or a small fee. Therefore, intrapreneurial librarianship is not focused primarily on generating profits but on encouraging an entrepreneurial mindset to inspire a fresh approach to library services.

Intrapreneurs are defined by Lemkau et al. (1991), those who rely primarily on innovation for success are ready to assume the associated risks and build a structure and environment within an organisation that embraces and supports innovation. Furthermore, intrapreneurs are independent, competitive, ambitious, individualistic, and motivated by problem-solving, bringing about change, innovating, boosting productivity, and attaining goals (Lemkau et al., 1991, p. 272). In the library world, librarian intrapreneurs have essential entrepreneurial skills, such as allocating resources for service quality, delegating and organising, reducing individual and team stress, accepting leadership responsibility, motivating at all levels, and selecting and developing a good team (Pantry & Griffiths, 2000). DeVries (2003) contends that almost every type of library has intrapreneurs, even though librarians may not match the stereotype of profit-driven, innovative risk-taker entrepreneurs.

Pantry and Griffiths (2000) asserted that the information and library service is suitable for the development of the entrepreneurial spirit. However, academic librarians require additional training to develop their understanding of entrepreneurship and intrapreneurship. Moreover, Lemkau et al. (1991) claimed that "intrapreneurship is an effective managerial style and problem-solving technique for information support units" (p. 275). However, there is a gap in the literature. There is a dearth of research into the viability of intrapreneurial approaches in academic libraries and the process for developing intrapreneurial activities. Moreover, there have yet to be empirical studies on intrapreneurial activities in Philippine academic libraries. Studies on entrepreneurial librarianship are primarily focused on discussion on entrepreneurial initiatives of academic and public libraries abroad, the status of entrepreneurship in libraries, entrepreneurial orientation and capabilities of librarians, entrepreneurship resources in libraries, entrepreneurial leadership behaviours of librarians, and the role of libraries in entrepreneurial university (Doulani et al., 2020; Edens & Malecki, 2020; Faulkner, 2018; Hoppenfeld & Malaf, 2015; Keshavarz, 2021; Mayombya et al., 2019; Mckeown, 2010; Sheshadri & Vijayakumar, 2018). Literature indicates that libraries have a long history of intrapreneurial activities. Several case studies on various entrepreneurial and intrapreneurial initiatives have been published. However, research has yet to be undertaken on how Filipino librarians and library administrators generate new and creative ideas and identify and capitalise on opportunities within the framework of sustainable development.

2 Research Problem and Objectives

Academic librarians can become intrapreneurs if they approach their services from a business perspective, strive to innovate and enhance their existing services (Gupta & Rubalcaba, 2022), or develop new, creative services. This can be accomplished by identifying, assessing, and exploiting hidden opportunities that generate value beyond the conventional constraints of available resources (Crumpton & Bird, 2019; DeVries, 2003). Lumley (2014) argues that academic librarians must adopt an entrepreneurial mindset as they continually develop fresh ideas for giving their students the knowledge and skills they need to become independent lifelong learners. However, academic librarians, particularly Filipino academic librarians, might not be aware of concepts and practices in entrepreneurial librarianship. They may already be pursuing intrapreneurial initiatives within their organisations without realising it. On the other hand, academic librarians who desire to foster an entrepreneurial spirit may lack the competencies to initiate activities and programs.

Using Matzembacher et al.'s (2019) sustainable entrepreneurial process model, this paper investigates how library management develops intrapreneurial activities in terms of idea generation, opportunity recognition, opportunity development, venture launch, and impact measurement. In addition, it explores the driving factors for successfully implementing intrapreneurial initiatives and determines the challenges and solutions library administrators face in meeting the goals of intrapreneurial activities. By knowing, examining, and implementing intrapreneurial principles in academic libraries, library management can identify new opportunities to create new services and programs to promote lifelong learning and information access. These strategies can assist them in adapting to the ever-evolving needs of their users and the community at large.

This research will contribute to the expanding body of literature on entrepreneurial librarianship by highlighting the development of entrepreneurial initiatives in academic libraries in the Philippines. The results will also help library management and school administration understand and recognise the factors that can drive sustainable intrapreneurship in academic libraries. It also attempts to fill a gap in the literature on developing an intrapreneurial university library in the Philippines by offering empirical data on the issues of sustainable entrepreneurship.

3 Literature Review

3.1 Entrepreneurship in Libraries

Hui-Chen et al. (2014) examined the entrepreneurial process to investigate how motivation, opportunity, and ability (MOA) influence entrepreneurial intentions "through personal attitude, subjective norms, and perceived behavioural control" (p. 727). Leyden and Link (2015) developed a theoretical model of the entrepreneurial process, wherein entrepreneurs form social networks, select an innovation to pursue, and then plan out a search strategy to determine how to bring that innovation to reality. The proposed entrepreneurial process model includes "innate entrepreneurial characteristics, creation of social network, search for desired innovation, and success or failure" (Leyden & Link, 2015, p. 476). The case study of Matzembacher et al. (2019) examined the prior experience and knowledge of entrepreneurs in generating ideas and identifying and exploiting "opportunities in the context of sustainable development" (p. 1). They discovered that entrepreneurs' knowledge of initiatives that aims to solve social and environmental problems, experiences, and skills were determinants of idea generation. Their findings proposed a sustainable entrepreneurial process flow with five stages: idea generation, opportunity recognition, opportunity development, venture launch, and positive impact. Mets (2020) has recently conceptualised a novel multi-dimensional process model of the entrepreneurial process, which is a modified version of the "functional stage model of the entrepreneurial process" (p. 698). His entrepreneurial process model has four stages: "propositions, idea development, concept development, and business development" (p. 698).

3.2 Entrepreneurship in Libraries

Several articles and essays examine the roles of libraries in promoting entrepreneurship and providing various support services for entrepreneurs. Schauder (1987) attempted to determine whether academic libraries can adopt entrepreneurship and whether they are suited to it. He identified the basic and innovative accomplishments of librarians. Basic successes of librarians include working effectively within the confines of their job, attaining effectiveness in deploying personnel, and advancing incrementally within one's job. The implementation of a new policy, creation of a new orientation or direction, the discovery of a unique opportunity, development of a new product or service, opening of a new market, creation of a new process, procedure, or technology for continued use, and implementation of structural changes are examples of innovative achievements (Schauder, 1987).

Mckeown (2010) emphasised that librarians play a crucial role in accessing information resources, advising students on the best strategies to start a business, and teaching them deductive reasoning skills when evidence is insufficient or conflicting. In addition, Restivo (2014) said that academic libraries provide library spaces that encourage creativity among entrepreneurs; as such, they might serve as innovation hubs on campuses where entrepreneurship is growing. Macdonald and Vanduinkerken (2015) highlighted the specific traditional roles in academic libraries and the use of diverse business models to build entrepreneurial culture. They argue that academic libraries must transform into "entrepreneurial organisations" whose strategic decisions are founded on business models rather than long-held conventions.

According to Christe et al. (2017), libraries are important but frequently overlooked components of the country's entrepreneurial landscape. They described libraries as digital, decentralised networks that give access to information resources and creative tools to innovators from all facets of society. Additionally, Christe et al. (2017) considered libraries crucial gateways for first-time independent entrepreneurs to receive state and federal assistance for small enterprises. Cole and Lysiak (2017) studied the role of business librarians in communicating the significance of extensive patent research at the early stages of the startup life cycle. Similarly, Elliott et al. (2017) investigated the contribution of the University of Arizona Library in the early stages of invention development in the commercialisation pipeline. Sheshadri and Vijayakumar (2018) analysed and described the vital role of library professionals in the success of entrepreneurs who were successful in launching their businesses. Ghosh (2019) argued that libraries are essential in setting the groundwork for entrepreneurs, inspiring them from the outset, and creating an environment that supports an entrepreneurial culture. Recently, Harris (2022) conducted interviews with librarians from selected Caribbean university libraries to examine their entrepreneurial practices and prospects and their perceptions of the entrepreneurial library. He concluded that there is a lack of unanimity among librarians over the definition of entrepreneurship.

3.3 Entrepreneurial Initiatives of Libraries

Inventing and initiating entrepreneurial activities is common for practising librarians who have grown innovative by necessity (Scanlon & Crumpton, 2011). Schauder (1987) discussed why libraries engage in entrepreneurial activities, including boosting their financial resources, enhancing professional experience, offering financial rewards to library personnel, and showcasing academic libraries' potential to support economic growth and meet community needs. He further claimed that entrepreneurship demands not just change but risky change. According to Schauder (1987), creating an entrepreneurial library necessitates a challenging trade-off between the flexibility required for innovators to thrive and the control needed to maintain a library operational in the short to medium term. Kilgour (1992) discussed the entrepreneurial opportunities of libraries in the 1990s. He identified nine outstanding entrepreneurial librarians that have made major innovative advances.

Neal (2001) developed a typology for entrepreneurial activity in academic libraries: "market-based, customised delivery and packaging of information; centre for research and design; innovative applications of technology; faculty, university, industry partnerships; e-commerce; lab for experimentation in teaching and learning; publishing ventures; and technology transfer" (Carpenter, 2012, p. 14). Kirk (2001) highlighted the entrepreneurial strategies of Sheridan Libraries, which possess business-attractive assets. Dempsey (2004) presented examples of library revenue streams, including used book sales, cafés, passport services, facility rentals, consulting, and fee-based corporate information services. The effects of open access on library operations are examined by Schmidt et al. and how entrepreneurial libraries discovered new methods to serve their customers in the emerging mixed open-access-traditional (MOA) environment.

Furthermore, university libraries successfully collaborated in developing and implementing conferences honouring librarianship's entrepreneurial spirit (Crowe et al., 2019; Ritchie-Baum et al., 2020; Scanlon & Crumpton, 2011). In 2009, the University Libraries of the University of North Carolina at Greensboro and the Z. Smith Reynolds Library of Wake Forest University co-hosted the first Conference for Entrepreneurial Librarians (Crowe et al., 2019). The conferences aimed to showcase how innovative and creative librarians are and how they develop new products and services. In some libraries, entrepreneurship is also a component of instruction in information literacy. Onaade (2012) examined the history and justification of information consultancy or information brokerage as a profession in Nigeria. Kirkwood and Evans (2012) described how Purdue University libraries collaborated with faculty to teach entrepreneurship and marketing courses, providing support through information literacy instruction. On the other hand, Carroll et al. (2019) looked into how information literacy training might be utilised to introduce undergraduate health sciences students to the medical entrepreneurship life cycle.

Academic libraries are increasingly repurposing their spaces to provide entrepreneurs with neutral, creative places that encourage entrepreneurial thinking, such as makerspaces, communal tables, collaboration spaces, green spaces, computer laboratories, and solo spaces (Bieraugel, 2019; Lumley, 2014; Nichols et al., 2017; Stover et al., 2019); thereby turning the libraries into one of the innovative hotspots on university campuses. Bieraugel (2019) reviewed several space assessment tools to determine whether libraries support entrepreneurial thinking. Aagaard and Arguello (2015) provided a practical guide for business librarians to evaluate their compliance with licensing agreements for electronic resources in entrepreneurship. In addition to the fee-based services offered by librarians, Hoppenfeld and Malaf (2015) discussed some resources and services made available to entrepreneurs, such as networking, outreach, business incubator support, intellectual property research support, and library instruction. Feldmann (2015) surveyed the 31 small business development centres to determine how they collaborate with libraries and librarians to assist entrepreneurs. Gupta et al. (2022) studied 50 academic librarians in Europe, Asia, America, Africa, and Australia to determine why librarians of entrepreneurial libraries used social networking sites to assist the business community in their market research. They concluded that entrepreneurial libraries are more likely to adopt new technologies smoothly.

3.4 Challenges in Entrepreneurial Libraries

Only some studies addressing the challenges of entrepreneurial libraries have been conducted (Feldmann, 2015; Hoppenfeld et al., 2013; Jusic, 2016; Mamza et al., 2021; Onaade, 2012; Schauder, 1987). For example, Schauder (1987) briefly discussed several barriers to entrepreneurial librarianship, such as performance-based library budget, multiple constituents, and "do good" ideology for public-service organisations. According to Onaade (2012), the tepid attitude of users towards librarians, the high illiteracy rate, and the lack of ICT infrastructure are obstacles faced by information brokers or independent information professionals in Nigeria. Hoppenfeld et al. (2013) stated that conducting market research, providing business literature, and maintaining market data databases can incur substantial expenditures. They also indicated that collaboration between librarians from other institutions increases the problem of insufficient research support staff time.

4 Research Methods

A qualitative multiple-case study method was used to collect data. A case study design is used when the purpose of the study is to address the "how" and "why" questions and to cover contextual conditions to describe a complicated real-world phenomenon (Yin, 2003, as cited in Baxter & Jack, 2015). The primary objective of this study is to investigate how library management develops intrapreneurial activities and explore the driving factors for successful intrapreneurial initiatives in academic libraries. According to Bouckenooghe et al. (2007, as cited in Bagheri & Pihie, 2010), "at the heart of entrepreneurship lie disjointed, discontinuous and non-linear events that cannot be studied with methods designed for continuous and linear processes" (p. 472). Thus, a case study design is suitable for process-related research (Hansen et al., 2012) such as this.

The subjects of this study were five librarians from four academic libraries that have received awards from the Philippine Association of Academic/Research Librarians (PAARL) for Outstanding Library of the Year, Outstanding Library Program of the Year, or Innovative Library Design Award. The academic libraries were selected because they have developed innovative and sustainable services and programs that other libraries and information centres can replicate. To preserve anonymity, the subjects were referred to as Library A, the advocator; Library B, the designer; Library C, the gamer; and Library D, the trendy.

Data was collected primarily through semi-structured interviews recorded using Zoom. The discussions focused on the development and implementation of entrepreneurial activities based on Matzembacher et al.'s (2019) framework, the identification of motivating factors that influence the success of entrepreneurial activities, and discussion on the challenges they face and solutions to address the challenges. The interviews were transcribed verbatim using the closed captions feature of Canvas Studio Basic. Then, the authors evaluated the transcriptions for completeness and accuracy and sent them to the participants for validation. The data were analysed in two phases: preliminary data analysis and thematic analysis.

5 Results and Discussions

5.1 Idea Generation

The findings suggest that the development of entrepreneurial activities initiated by library management has been a dynamic and iterative process that continues to evolve. As stated by Ge et al. (2022), citing Vogel (2017), the entrepreneurial process can be viewed as a progression from the initial concept to the opportunity for a new venture and expansion. The transition between the generation of ideas and the final implementation of an entrepreneurial opportunity takes time. In the early stages of the entrepreneurial process, ideas are stimulated and generated (McMullen & Shepherd, 2006). For example, Library B narrated that the idea of creating collaborative spaces started when the university president asked the library director whether she wanted to have a renovated library or a new library building. This strongly shows that top management can influence the development of new services and products by fostering an environment that encourages creativity inside the organisation (Chen, 2007). Therefore, as McMullen and Shepherd (2006) mentioned, individual and external system-level factors shape ideation activities.

Furthermore, the participants reported that the library director, head librarians, librarians, and support staff were involved during the brainstorming process. They also claimed that, both directly and indirectly, library users participated in the ideation process through customer feedback and user experience surveys. Since the lead librarians of Library C are also students at the same university, they were able to put themselves in the shoes of the library users who would be most interested in their gamification activity while brainstorming. Likewise, Library B reviewed the longstanding customer feedback and discussed which concerns could be addressed by the new learning spaces. According to Desouza (2017), managing ideas that do not consider clients is challenging but possible. He added that disclosing problems of interest to employees and soliciting their input is also essential. In generating fresh ideas, the participants utilised brainstorming and library benchmarking. Because not all ideas are inherently innovative and creative (Ge et al., 2022), they engage in brainstorming. Two participants reported participating in local and international learning visits. On the other hand, Library C examined the other libraries' websites and social media accounts to determine what other libraries do to increase user involvement in library instruction and look into their marketing strategies.

Moreover, the participants claim that idea-sharing in the workplace is highly encouraged. Specifically, Library D shared that staff meetings provide opportunities to hear frontline staff feedback and discuss challenges and solutions.

5.2 Opportunity Recognition and Development

After generating an idea, intrapreneurs must take additional steps to identify the optimal opportunity to build a new venture, service, or product (Ge et al., 2022). The results demonstrate that librarians' expertise, desire to innovate and rebrand, dissatisfaction with existing services and programs, the availability of library resources, and customers' feedback can lead to opportunity recognition and exploitation. Herron and Sapienza (1992, as cited in Shao et al., 2020) claimed that unacceptable dissatisfaction drives individuals to seek new opportunities. Shao et al. (2020) added that the entrepreneur's competencies, past accomplishments, and personality subconsciously affect the search for ideas and the level of aspiration. In the opportunity development stage, Library D reported that "*Some ideas needed further exploration. Although the purpose is clear, other things need to be considered, such as timeline, staffing, logistics, and budget… More research needed to be done, and a structured proposal had to be drafted.*" Dimov (2007) supports such action. He defined opportunity development as the growth or shaping of an idea, which entails a continual stream of ideas (Dimov, 2007). de Koning (2003, as cited in Hansen et al., 2012) said that cognitive processes involved in generating ideas, including thinking-through-talking, can assist entrepreneurs in realising that their opportunities require further development if they have difficulty articulating them.

Ideas can be evolved into numerous forms, such as business concepts, business models, business plans, and new businesses (Hansen et al., 2012). In the opportunity development stage, the participants of this study created project proposals or plans based on the idea, outlining the purpose, target recipients, guidelines, mechanics, logistics, technology requirements, staffing, budget, and evaluation. The proposals were presented to the employees, library management, and school administrators for evaluation, feedback, approval, and support. The participants stated that proposals are modified or altered suitably to meet customers' needs and demands. According to Matzembacher et al. (2019), assessing the social needs and assets associated with opportunity recognition is necessary. Library D emphasised the importance of conducting research "*to check available technologies and processes to deliver the service, and to determine buy-in from concerned offices.*" During this phase, key library personnel participated in the planning. Library C reported that its strategic partners, such as publishers and service providers, were involved in creating materials and narratives for the gamification activity. When participants were asked when they realised that the idea could be executed and that the opportunity to do so exists, their responses varied. Library D responded

> *when it responded to an existing or impending challenge, when it provided an opportunity to insert innovation and creativity into processes and programs that we have been doing or have not been doing, and when it presented the possibility of the library being the "first" to do it.*

Hansen et al. (2012), citing Stevenson and Gumpert (1985), stated that opportunities are at the heart of entrepreneurship and that recognising and capitalising on opportunities are a successful entrepreneur's most crucial skills. Additionally, Sa'ari and Johare (2014) concluded that recognising opportunities is one of the entrepreneurial skills necessary for implementing innovative behaviour in academic libraries. Hence, academic librarians must develop their opportunity recognition skills.

5.3 Launching and Impact Measurement

Launching happens when new library services, spaces, and programs are announced and made available to the target market. It requires the development of strategies, the acquisition of any lacking tangible and intangible resources, and the organisation of a team (Matzembacher et al., 2019). Plans for launching intrapreneurial activity vary. Library D said, "*depending on the activity, the launch could be an email communication with announcements through onsite posters, videos, and social media or a formal launching program with a walk-through, exhibit, or activities.*" Library B said that the launch of the new library spaces was held during the 70th founding anniversary of the university. In the case of Library C, the launch was part of a bigger event. The participants utilised various resources in the launch, such as electronic bulletins, videos, and staff skills. Two libraries in this study reported collaborating with colleges and other offices when necessary. Some participants reported mistakes or failures during the implementation, such as failure to provide workstations for librarians, miscommunication of the plans and game mechanics, and late furniture delivery. Since intrapreneurial activity involves taking risks and testing new ideas, tolerance for mistakes and failures is necessary (Ng, 2012). Therefore, library management should not punish failures but view them as learning opportunities (Ng, 2012).

It is essential to evaluate the effectiveness of intrapreneurial endeavours based on key performance indicators that emphasise outcomes and impacts (Appleton, 2018). When the participants were asked about the outcomes of their intrapreneurial efforts, their responses were positive. After the launch, Library B sought increased social media engagement and received numerous requests from alumni, school administrators, and other libraries to visit the new building. Library C observed a utilisation increase of some databases. In addition, Library A highlighted the encouraging feedback from students and faculty members who used the virtual reference services amid the COVID-19 outbreak. As such, the participants measured their success based on resource utilisation, user satisfaction, customer feedback, and return experience. Furthermore, their accomplishments were recognised with mini celebrations for the library staff, highlighting victories during staff meetings and in the annual report, releasing posters and infographics, and honouring staff members directly involved in the program. According to Pantry and Griffiths (2000), intrapreneurial libraries must acknowledge their successes and disseminate information regarding their accomplishments within the organisation. They added that critical success factors (CFS) must be developed for each service, and there must be an ongoing conversation between libraries and their patrons to provide the necessary feedback. Appleton (2018) cited some examples of CFS for libraries as follows: "efficient and reliable suppliers; motivated, skilled and technically expert staff; accessible service models; a robust IT network infrastructure; and customer-focused mission" (par. 6).

5.4 Motivation and Success Factors

When participants were asked what motivated them to produce ideas and transform them into a new service or program, their responses differed. For instance, the director of Library A explained that the circumstances when the COVID-19 pandemic struck in 2020 compelled her to think outside the box strategies to reach students and faculty remotely, such as virtual reference services via Facebook Messenger, and she had to persuade school administrators that librarians and support staff are a necessary workforce to prevent layoffs. Similar to Sarma et al.'s (2022) findings, the fundamental premise of opportunity recognition or new service creation was the introduction of new services, new approaches, and new communication channels for students and faculty. In addition, Library B and Library C were motivated by the decision of top management, the desire to address the customers' feedback, user satisfaction, the desire to rebrand the image of the library profession, the declining attendance in information literacy webinars, and the personal vision to do better or create new services. Recognising a solution to a problem presents an opportunity (Matzembacher et al., 2019) to create new or enhance existing library services, programs, and spaces. According to Library D, *"foresight is important in generating ideas because it allows the library to be ready and proactive in responding to the users' needs. Other enablers include user needs, support from administration, technology, staff expertise, and budget."*

Furthermore, the participants were asked to identify the factors leading to the success of intrapreneurial efforts. Among the factors contributing to successful intrapreneurship in libraries are innovative culture, strong collaboration with student organisations, effective communication skills, extensive and careful planning, customer insights, stakeholder support, staff expertise, technology, budget, and partnerships relationships. On the other hand, lack of collaboration, communication, foresight, and planning could be among the factors for failure. These findings are consistent with studies of Yeh and Ramirez (2016), Antoncic and Hisrich (2001), Kitsios and Grigoroudis (2020), Häikiö and Koivumäki (2016), Wang et al. (2022), and Sung et al. (2011). Specifically, Huang et al. (2021) categorised enablers of intrapreneurship as enablers-individual level (e.g., self-attitude, capabilities, judgments, and personality attributes) and enablers-organizational level (e.g., developmental support, work design, managerial style, and innovative culture). According to Sa'ari and Johare (2014), the availability of technology and a company's capacity to leverage that technology contributed significantly to innovative performance.

5.5 Challenges Along the Way

Executing entrepreneurial activities entails various risks, difficulties, and challenges (Ahmed & Harrison, 2021). As a challenge, Library A noted changing administrators and unsupportive university administrators. There were delays in the delivery of library furnishings to Library B. Library C indicated that obtaining participants from various autonomous campus units, communicating with other university libraries in the system, and accommodating staff schedules were challenging. Due to their lack of knowledge, Library C participants struggled to construct the gamified library orientation. Limited budget and technology become challenges for Library D. Matzembacher et al. (2019) agreed with this finding that funding and other forms of support for entrepreneurial activity are also obstacles. In addition, the attitude of employees toward introducing new services may also be an impediment, particularly when employees perceive new services as an additional workload. Mayombya et al. (2019) also discovered that funding to initiate and sustain intrapreneurial ventures in the library is challenging. Leitch and Volery (2017) argued that entrepreneurial leaders' personal and functional competencies are essential for overcoming obstacles and dealing with ongoing innovation and change.

6 Conclusion

This study's primary purpose is to analyse how library management creates intrapreneurial activities and identify the factors motivating successful intrapreneurial efforts in academic libraries. The findings suggest that the intrapreneurial process, from ideation to impact measurement, is dynamic and iterative. Developing intrapreneurial activities requires commitment, dedication, and a willingness to learn from successes and failures. Moreover, the findings indicate that creating new or modifying existing library services and programs to enhance the user experience is the key motivator for innovation. Employee involvement and customer feedback are essential to successful idea generation, opportunity recognition, and opportunity exploitation. Finally, the participants demonstrated positive changes in user behaviour and resource utilisation resulting from intrapreneurial endeavours. Although this study did not examine whether the intrapreneurial behaviours of the participants contributed to the successful implementation of intrapreneurial activities, it is recommended that academic libraries adopt an intrapreneurial mindset and cultivate a culture of innovation, creativity, and risk-taking. Academic libraries can create new services, programs, and places to satisfy the changing demands of their communities if they nurture an intrapreneurial culture. Encouraging employees to innovate can drive organisational growth and help academic libraries develop a sustainable approach to achieve innovations within the library and contribute to the library's competitiveness (Alam et al., 2020). Therefore, the benefits of intrapreneurship in academic libraries, if they employ it effectively, include remaining relevant, adapting to changing environments, driving growth, staying competitive, and developing a sustainable approach to achieve innovations (Alam et al., 2020; Huang et al., 2021; Moghaddas et al., 2020; Vassilakopoulou & Grisot, 2020).

References

Aagaard, P., Arguello, N.Z.: Practical approaches to compliance for entrepreneurial uses of licensed databases in libraries. Ref. Serv. Rev. **43**(3), 419–438 (2015). https://doi.org/10.1108/RSR-03-2015-0016

Ahmed, F., Harrison, C.: Challenges and competencies of entrepreneurial leaders in driving innovation at DIY laboratories. Technol. Anal. Strateg. Manag. **33**(10), 1132–1146 (2021). https://doi.org/10.1080/09537325.2021.1908538

Alam, M.Z., Nasir, N., Rehman, C.A.: Intrapreneurship concepts for engineers: a systematic review of the literature on its theoretical foundations and agenda for future research. J. Innov. Entrep. **9**(1) (2020). https://doi.org/10.1186/s13731-020-00119-3

Antoncic, B., Hisrich, R.D.: Intrapreneurship: construct refinement and cross-cultural validation. J. Bus. Ventur. **16**(5), 495–527 (2001). https://doi.org/10.1016/S0883-9026(99)00054-3

Antoncic, J.A., Antoncic, B.: Employee satisfaction, intrapreneurship and firm growth: a model. Ind. Manag. Data Syst. **111**(4), 599–607 (2011). https://doi.org/10.1108/02635571111133560

Appleton, L.: Using key performance indicators to measure library performance. Library Connect (2018, April 4). https://libraryconnect.elsevier.com/articles/using-key-performance-indicators-measure-library-performance

Bagheri, A., Pihie, Z.A.L.: Entrepreneurial leadership learning: in search of missing links. Procedia Soc. Behav. Sci. **7**(2), 470–479 (2010). https://doi.org/10.1016/j.sbspro.2010.10.064

Baxter, P., Jack, S.: Qualitative case study methodology: study design and implementation for novice researchers. Qual. Rep. **13**(4), 544–559 (2015). https://doi.org/10.46743/2160-3715/2008.1573

Bieraugel, M.: Do your library spaces help entrepreneurs?: space planning for boosting creative thinking. Adv. Libr. Admin. Organ. **40**, 21–32 (2019). https://doi.org/10.1108/S0732-067120190000040001

Carpenter, M.T.H.: Cheerleader, opportunity seeker, and master strategist: ARL directors as entrepreneurial leaders. Coll. Res. Libr. **73**(1), 11–32 (2012). https://doi.org/10.5860/crl-192

Carroll, A.J., Hallman, S.J., Umstead, K.A., McCall, J., DiMeo, A.J.: Using information literacy to teach medical entrepreneurship and health care economics. J. Med. Libr. Assoc. **107**(2), 163–171 (2019). https://doi.org/10.5195/jmla.2019.577

Chen, M.H.: Entrepreneurial leadership and new ventures: creativity in entrepreneurial teams. Creativ. Innov. Manag. **16**(3), 239–249 (2007). https://doi.org/10.1111/j.1467-8691.2007.00439.x

Christe, D., Bhatt, J.J., McGee, D.G., Wolfish, R.: Entrepreneurship, engineering, innovation, and libraries: empowering innovators with information. In: ASEE Annual Conference and Exposition, Conference Proceedings, p. 18444 (2017). https://doi.org/10.18260/1-2--28290

Cole, C., Lysiak, L.: Due diligence: recognizing the role of patent research in the start-up life cycle. J. Bus. Financ. Librariansh. **22**(3–4), 222–230 (2017). https://doi.org/10.1080/08963568.2017.1372014

Crowe, K., Crumpton, M.A., Gwynn, D., Harper, J., Lock, M.B., Scanlon, M.G.: The status of entrepreneurship in libraries: content analysis and assessment from the conference for entrepreneurial librarians. Adv. Libr. Admin. Organ. **40**, 115–132 (2019). https://doi.org/10.1108/S0732-067120190000040007

Crumpton, M.A., Bird, N.J.: Educating the entrepreneurial librarian. Adv. Libr. Admin. Organ. **40**, 169–182 (2019). https://doi.org/10.1108/S0732-067120190000040011

Dempsey, B.B.: Cashing in on service: entrepreneurial ventures make money and extend library's mission. Libr. J. **129**(18), 38–41 (2004)

Desouza, K.V.: Intrapreneurship: Managing Ideas Within Your Organization. University of Toronto Press (2017). http://linkinghub.elsevier.com/retrieve/pii/S0040162513003223

DeVries, J.: Entrepreneurial librarians: embracing innovation and motivation. Sci. Technol. Libr. **24**(2–3), 209–217 (2003). https://doi.org/10.1300/J122v24n01_14

Dimov, D.: Beyond the single-person, single-insight attribution in understanding entrepreneurial opportunities. Entrep. Theory Pract. **31**(5), 713–731 (2007). https://doi.org/10.1111/j.1540-6520.2007.00196.x

Doulani, A., Sahebi, S., Saberi, M.K.: Assessing the entrepreneurial orientation of the librarians: a case study of Iranian public libraries. Glob. Knowl. Mem. Commun. **69**(4–5), 253–268 (2020). https://doi.org/10.1108/GKMC-10-2019-0120

Edens, W., Malecki, A.L.: Entrepreneurship initiatives in academic libraries. Public Serv. Q. **16**(2), 107–111 (2020). https://doi.org/10.1080/15228959.2020.1736707

Elliott, C., Dewland, J., Martin, J.R., Kramer, S., Jackson, J.J.: Collaborate and innovate: the impact of academic librarians on the commercialization of university technology. J. Libr. Adm. **57**(1), 36–48 (2017). https://doi.org/10.1080/01930826.2016.1215674

Faulkner, A.E.: Entrepreneurship resources in US public libraries: website analysis. Ref. Serv. Rev. **46**(1), 69–90 (2018). https://doi.org/10.1108/RSR-07-2017-0025

Feldmann, L.M.: Small business development centers and libraries: a survey. Ref. Serv. Rev. **43**(3), 369–378 (2015). https://doi.org/10.1108/RSR-06-2014-0016

Franco, M., Pinto, J.: Intrapreneurship practices in municipal archives: a practice-oriented study. J. Librariansh. Inf. Sci. **49**(2), 165–179 (2017). https://doi.org/10.1177/0961000616632055

Ge, B., Wang, Q., Yao, M.: From ideas to entrepreneurial opportunity: a study on AI. Syst. Res. Behav. Sci. **39**(3), 618–632 (2022). https://doi.org/10.1002/sres.2874

Ghosh, M.: Global trend in entrepreneurial librarianship: a review of publications and best practices [Paper presentation]. IFLA World Library and Information Congress 2019, pp. 1–11 (2019). http://library.ifla.org/id/eprint/2844

Gupta, V., Rubalcaba, L.: University libraries as open innovation partners: harnessing hidden potential to foster global entrepreneurship. J. Acad. Librariansh. **48**(2), 102432 (2022). https://doi.org/10.1016/j.acalib.2021.102432

Gupta, V., Rubalcaba, L., Gupta, C., Pereira, L.: Social networking sites adoption among entrepreneurial librarians for globalizing startup business operations. Library Hi Tech. (2022). https://doi.org/10.1108/LHT-05-2022-0234

Häikiö, J., Koivumäki, T.: Exploring digital service innovation process through value creation. J. Innov. Manag. **4**(2), 96–124 (2016). https://doi.org/10.24840/2183-0606_004.002_0006

Hansen, D.J., Mcmurchie, L., Monllor, J.: Opportunity development: an exploratory study of ecopreneurs using a creativity perspective. J. Res. Mark. Entrep. **14**(1), 27–39 (2012). https://doi.org/10.1108/14715201211246733

Harris, S. (2022). Building an entrepreneurial Caribbean university library: practice, possibilities, and lessons. Alexandria: J. Natl. Int. Libr. Inf. Issues 1–27. https://doi.org/10.1177/09557490221103034

Hoppenfeld, J., Malaf, E.: Engaging with entrepreneurs in academic and public libraries. Ref. Serv. Rev. **43**(3), 379–399 (2015). https://doi.org/10.1108/RSR-02-2015-0011

Hoppenfeld, J., Wyckoff, T., Henson, J.A.J., Mayotte, J.N., Kirkwood, H.P.: Librarians and the entrepreneurship Bootcamp for veterans: helping disabled veterans with business research. J. Bus. Financ. Librariansh. **18**(4), 293–308 (2013). https://doi.org/10.1080/08963568.2013.825227

Huang, L.Y., Yang Lin, S.M., Hsieh, Y.J.: Cultivation of intrapreneurship: a framework and challenges. Front. Psychol. **12** (2021). https://doi.org/10.3389/fpsyg.2021.731990

Hui-Chen, C., Kuen-Hung, T., Chen-Yi, P.: The entrepreneurial process: an integrated model. Int. Entrep. Manag. J. **10**(4), 727–745 (2014). https://doi.org/10.1007/s11365-014-0305-8

Jusic, J.: What happens when entrepreneurial public libraries change directors? [Simmons College] (2016). https://www.proquest.com/docview/1954036048

Keshavarz, H.: Entrepreneurial capabilities of librarians in university libraries: a cross-contextual study on the impact of information literacy. J. Bus. Financ. Librariansh. **26**(3–4), 200–222 (2021). https://doi.org/10.1080/08963568.2021.1941576

Kilgour, F.G.: Entrepreneurial leadership. Libr. Trends **40**(3), 457–474 (1992). https://core.ac.uk/download/pdf/4817121.pdf

Kirk, E.E.: The entrepreneurial library: creating revenue in the age of e-commerce. In: Crossing the Divide: Proceedings of the Tenth National Conference of the Association of College and Research Libraries, vol. 10, pp. 255–258 (2001)

Kirkwood, H., Evans, K.: Embedded librarianship and virtual environments in entrepreneurship information literacy: a case study. J. Bus. Financ. Librariansh. **17**(1), 106–116 (2012). https://doi.org/10.1080/08963568.2011.630583

Kitsios, F.C., Grigoroudis, E.: Evaluating service innovation and business performance in tourism: a multicriteria decision analysis approach. Manag. Decis. **58**(11), 2429–2453 (2020). https://doi.org/10.1108/MD-09-2019-1326

Lemkau, H.L., Burrows, S., Stolz, F.: An intrapreneurial approach to integrating information production services into an academic medical center library. Bull. Med. Libr. Assoc. **79**(3), 271–275 (1991)

Leyden, D.P., Link, A.N.: Toward a theory of the entrepreneurial process. Small Bus. Econ. **44**(3), 475–484 (2015). https://doi.org/10.1007/s11187-014-9606-0

Lumley, R.M.: A coworking project in the campus library: supporting and modeling entrepreneurial activity in the academic library. New Rev. Acad. Librariansh. **20**(1), 49–65 (2014). https://doi.org/10.1080/13614533.2013.850101

Macdonald, K.I., Vanduinkerken, W.: Libraries surviving as entrepreneurial organizations: a creative destruction perspective. New Libr. World **116**(7–8), 406–419 (2015). https://doi.org/10.1108/NLW-01-2015-0005

Mamza, W.P., Waba, Y.Y., State, Y.: Towards a sustainable entrepreneurial librarianship: problems and approaches for innovative achievement. Int. J. Manag. Stud. Bus. Entrep. Res. **6**(1), 32–44 (2021)

Matzembacher, D.E., Raudsaar, M., de Barcellos, M.D., Mets, T.: Sustainable entrepreneurial process: from idea generation to impact measurement. Sustainability (Switzerland) **11**(21), 1–26 (2019). https://doi.org/10.3390/su11215892

Mayombya, J.E., Mwantimwa, K., Ndenje-Sichalwe, E.: Entrepreneurial opportunities: a roadmap for diversifying financial sources in libraries, Tanzania. IFLA J. **45**(4), 322–333 (2019). https://doi.org/10.1177/0340035219868824

Mckeown, K.R.: Challenges and solutions for libraries in serving entrepreneurship needs: findings from ProQuest research. J. Bus. Financ. Librariansh. **15**(3–4), 253–260 (2010). https://doi.org/10.1080/08963568.2010.488048

McMullen, J.S., Shepherd, D.A.: Entrepreneurial action and the role of uncertainty in the theory of the entrepreneur. Acad. Manag. Rev. **31**(1), 132–152 (2006). https://doi.org/10.5465/amr.2006.19379628

Mets, T.: Exploring the model of the entrepreneurial process. In: Australian Centre for Entrepreneurship Research Exchange Conference 2015. Conference Proceedings, Queensland University of Technology, Adelaide, October, pp. 693–707 (2020)

Mezeh, A.A., Bassey, U.L., Adim, C.V.: Entrepreneurial leadership: a review of literature. Int. Acad. J. Manag. Mark. Entrep. Stud. **8**(2), 210–221 (2021). http://www.arcnjournals.org/images/ASA-IAJMMES-8-3-110.pdf

Moghaddas, S.Z., Tajafari, M., Nowkarizi, M.: Organizational empowerment: a vital step toward intrapreneurship. J. Librariansh. Inf. Sci. **52**(2), 529–540 (2020). https://doi.org/10.1177/0961000619841658

Neal, J.G.: The entrepreneurial imperative: advancing from incremental tb radical change in the academic library. Portal: Libr. Acad. **1**(1), 1–13 (2001). https://doi.org/10.1353/pla.2001.0006

Ng, J.C.Y.: Implementing Intrapreneurship: a structural and cultural approach. Graziadio Bus. Rev. **15**(3) (2012)

Nichols, J., Melo, M.M., Dewland, J.: Unifying space and service for makers, entrepreneurs, and digital scholars. Portal: Libr. Acad. **17**(2), 363–374 (2017). https://doi.org/10.1353/pla.2017.0022

Onaade, O.J.: Information brokerage: an entrepreneurial approach to information services in Nigeria. Glob. J. Manag. Bus. Res. **12**(20), 13–19 (2012)

Pantry, S., Griffiths, P.: Being an intrapreneur and creating a successful information service within your organization. Bus. Inf. Rev. **17**(4), 205–214 (2000). https://doi.org/10.1177/0266382004237773

Restivo, L.: Why startups need libraries (and librarians). Ser. Libr. **67**(1), 31–37 (2014). https://doi.org/10.1080/0361526X.2014.915610

Ritchie-Baum, M., Thynne, S.M., Cramer, S.M.: Creating the entrepreneurship & libraries conference 2020: a collaboration of public, special, and academic librarians, vendors, and economic development stakeholders. Collab. Librariansh. **12**(3/4), 231–241 (2020)

Sa'ari, H., Johare, R.: Leveraging entrepreneurial knowledge competencies among academic librarians in Malaysia. J. Inf. Knowl. Manag. **4**(1), 79–95 (2014)

Sarma, S., Attaran, S., Attaran, M.: Sustainable entrepreneurship: factors influencing opportunity recognition and exploitation. Int. J. Entrep. Innov. (2022). https://doi.org/10.1177/14657503221093007

Scanlon, M., Crumpton, M.: Re-conceiving entrepreneurship for libraries: collaboration and the anatomy of a conference. Collab. Librariansh. **3**(1), 16–27 (2011). https://doi.org/10.29087/2011.3.1.07

Schauder, D.: Entrepreneurship and the academic library: insights from organization theory. IATUL Q. **2**(1), 36–50 (1987)

Shao, Q.G., Liou, J.J.H., Weng, S.S., Su, P.: Constructing an entrepreneurship project evaluation system using a hybrid model. J. Bus. Econ. Manag. **21**(5), 1329–1349 (2020). https://doi.org/10.3846/jbem.2020.13165

Sheshadri, K.N., Vijayakumar, S.: Role of libraries in entrepreneurship and social change: review of selected case studies. Int. J. Recent Trends Bus. Tour. **2**(4), 63–67 (2018)

Stover, M., Jefferson, C., Santos, I.: Innovation and creativity: a new facet of the traditional mission for university libraries. Adv. Libr. Admin. Organ. **40**, 135–151 (2019). https://doi.org/10.1108/S0732-067120190000040006

Sung, S.Y., Cho, D.-S., Choi, J.N.: Who initiates and who multi-agent model of organizational innovation. J. Manag. Organ. **17**(3), 344–363 (2011)

Vassilakopoulou, P., Grisot, M.: Effectual tactics in digital intrapreneurship: a process model. J. Strat. Inf. Syst. **29**(3), 1–24 (2020). https://doi.org/10.1016/j.jsis.2020.101617

Wang, Q., Oshri, I., Zhao, X.: Value cocreation in new service development: a process-based view of resource dependency. Eur. J. Mark. **56**(1), 184–208 (2022). https://doi.org/10.1108/EJM-04-2020-0319

Yeh, S.T., Ramirez, R.: A conceptual model of service innovation: the case of academic libraries. In: 22nd Americas Conference on Information Systems, pp. 1–5 (2016)

Literacies

Promoting Academic Writing Literacy by Training Programs Evaluation Based on the One-Group Pre-test/Post-test Design

Cheng Bing[1(✉)], Li Ling[1,2], Liu Yanan[2], and Du Jiangyuan[3]

[1] National Science Library, Chinese Academy of Sciences, Beijing, China
chengbing@mail.las.ac.cn
[2] University of Chinese Academy of Sciences, Beijing, China
[3] Shanghai Normal University, Shanghai, China

Abstract. *Purpose*: Writing research papers for publication in international academic journals is necessary and important for researchers. Almost all researchers have to practice academic writing skills during their research career. Specific academic writing education programs are being developed in various academic libraries to promote academic writing. One online short-term summer training program was designed for this purpose which has attracted researchers to register for the training. This study aims to evaluate the effect on improving the writing skills for researchers attended the short-term training course. *Methods*: The pre-test/post-test design (PPD), one of the most frequently used experimental methods and particularly popular in educational research, is used in this study. The effective sample was composed of 574 researchers who completed the self-assessment questionnaires. Totally 17 questions were designed to evaluate the effects. The SPSS statistical analysis was used to analyze the data. *Results and Conclusion*: The results show that researchers improved their writing skills through the short-term training, which show significant difference between the experimental and the control groups. It also shows difference between the researchers with different characteristics, studied in different stages, ever published English academic articles, or ever learned the writing curriculum before. These results are useful for an academic library to explore educational activities on academic writing. This study provides pioneering evidence for testing the effectiveness of the training on academic writing on promoting the writing skills for English-as-a-second-language (ESL) researchers.

Keywords: Academic writing skill · Training program · Pre-test/post-test design · English as a second language (ESL)

1 Introduction

Academic writing is vital for researchers for their scholarly publishing and communications. The ability to write clearly, concisely and logically is critical to academic success. However, academic writing can be a challenging task for researchers, especially for

A. Katsirikou (Ed.): QQML 2023, 68, pp. 37–48, 2026.
https://doi.org/10.1007/978-3-032-01264-7_3

those whose English is a second language. Their research careers often relies on their ability to publish their research in international peer-reviewed journals. Therefore, it is important to provide them with support and training to improve their academic writing abilities.

Academic libraries play a crucial role in helping researchers develop their academic writing literacy. According to the Association of College and Research Libraries (ACRL), the guideline of research library was provided for general education courses in Framework for Information Literacy for Higher Education (ACRL 2023). It is also full of challenges to integrate into specific curricula in graduate education. Writing literacy is the complex information competency from the information to create the knowledges.

In universities, academic writing curricula are provided for graduate students in the first or second semester. The contents are general and include common writing types, such as experimental reports, essays, research proposals, thesis and dissertations, and research papers, etc. However, the curricula which focus on how to write academic journal articles are relatively fewer. And the guidance is insufficiently embedded with scenario-based writing.

A 3.5-day short-term online summer academic writing training program was thus designed for improving research paper writing for international academic journal publications and conducted by the National Science Library of the Chinese Academy of Sciences (NSL) through our online virtual conference platform, which is a good chance for researchers to plan and write papers and has attracted a lot of them to register for the training.

Regarding the effect of the course on the participants, however, because academic writing is a demanding process that involves a number of specialized skills, it is hard to make accurate assessment on it. Some methods have been developed to evaluate the impacts of face-to-face courses, such as the writing process and feedback model (Carol 2001). But for the online learning courses, because of the pandemic, which deliver training contents through a virtual platform, the evaluation is more difficult and influenced by various factors.

In this study, we used a rapid evaluation approach, i.e. the one-group pre-test/post-test design (Emma and Carole 2012), to conduct such evaluation. A series of self-assessment questionnaires were applied to the researchers participated in the short-term online summer writing training program to evaluate the training effects on improving their writing levels and skills.

2 Methods

2.1 Design

The online short-term summer academic writing training program was conducted for 3.5 days, focusing on writing English research articles according to the overall publication requirements of international academic journals.

A one-group pre-test/post-test experiment study was conducted during the training program, which is a self-assessment questionnaire with the Likert 5-point scales including the options for the writing literacy with 1 to 5 scores representing the weakest to the strongest. Before the survey, a pilot one was first tested with a smaller sample.

2.2 Participants

Only graduate or doctoral students/candidates attended the whole academic writing training program who are interested in the survey were recruited as the volunteer participants of this study, i.e. the experimental group. Other researchers attended the program such as postdoctors, staff, associate professors, or other junior or senior researchers, etc. were excluded due to small sample size.

The control group involved the participants who did not attend the training course. Both groups took a pre-test before the course begins. After the training was completed, both two groups took a post-test in a week.

Others either did not attend the whole course or the pre-/post-test sessions were excluded as well.

2.3 Intervention

The training intervention was the online short-term summer writing training program in the form of lectures and workshops in the daytime. After the daily course, the participants wrote the gains or practiced the homework on the learning platform in spare time. The contents of the course are based on the journal research articles, covering the writing knowledge and skills in general. The instructors of training included English-speaking training experts, researchers in natural science, scholarship librarians, and writing tool application and publishing experts.

2.4 Data Collection

A total number of 5697 participants were registered the 3.5-day short-term online summer writing training program, among whom 5106 were excluded for the reasons that 2142 only attended part of the course, 2889 were absent from answering the self-assessment questionnaire, and other 75 participants are not graduate or doctoral students, making up the sample size of the experimental group as 591. Among the 591 selected participants, 17 were excluded further due to their extreme answers, such as giving all the same answers or finishing answering the questionnaire in a too short time. The effective questionnaires received were 574.

For writing literacy measurement, the self-assessment questionnaire was distributed twice before the first day (pre-test) and after the last day of the course (post-test) with at most 4 to 10 days interval. The questionnaire comprised 17 questions related to the fundamental writing aspects of research article writing and a 5-point Likert scale was used in the assessment. These questions were further divided into four sections: writing awareness (Awareness), basic writing knowledge (Knowledge A), advanced writing knowledge (Knowledge B), and writing tool application skills.

In the first section, the writing awareness self-assessment included 5 questions about writing plan, daily accumulation, logic thinking, writing style, and review writing habit. In the second section, the basic writing knowledge self-assessment included 4 questions about how to write the title, keywords, acknowledgements, and methods, which are the components of a journal paper.

In the third section, the advanced writing knowledge self-assessment included 5 questions of writing knowledge related to writing the abstract, introduction, result and/or discussion, and conclusion which are the key parts of a journal paper. In the fourth section, the writing tool application skills section included 3 questions related to table/figure drawing, managing notes or references, etc. The assessment questionnaire was designed to focus closely on the contents of the training course (Fig. 1).

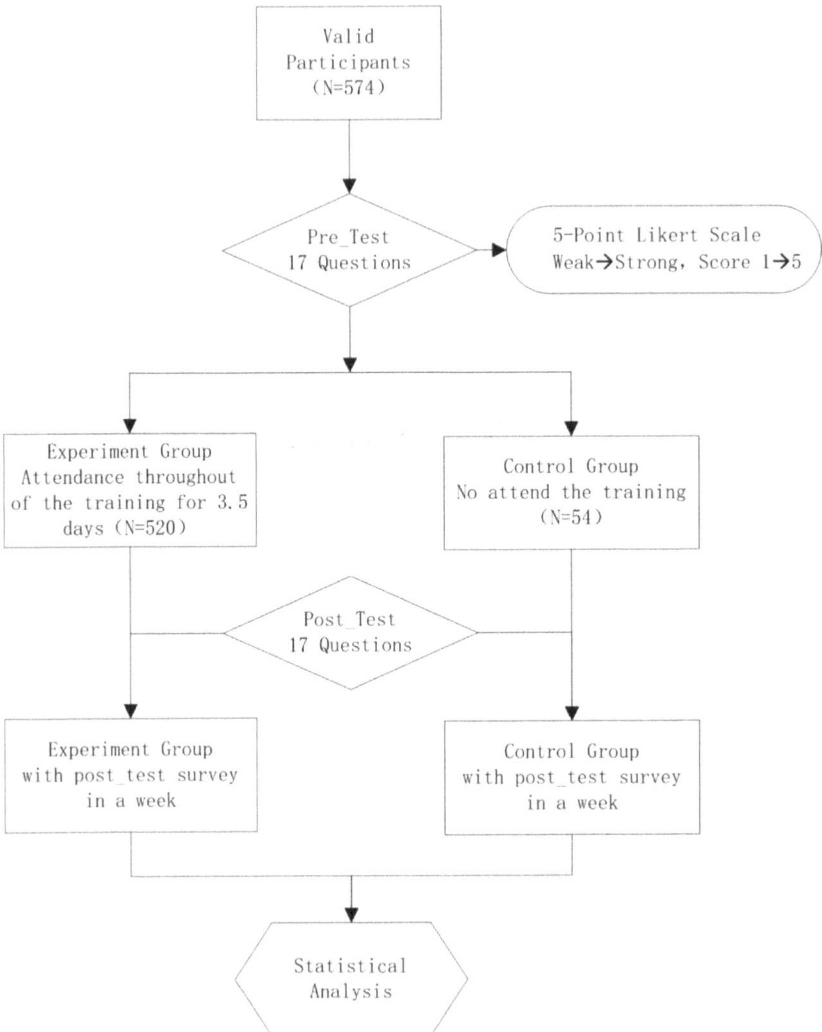

Fig. 1. The flow-chart of the training intervention.

3 Statistical Analysis

The SPSS version 26.0 statistical software (IBM Corporation, Armonk, NY, US) was used for data analysis. The general characteristics and degrees of study variables were the descriptive statistics compiled. Whereas the continuous variable of the pre-test and the post-test scores of the training course was expressed as mean ± *S.D.*

The independent sample t-test/analysis of variance (ANOVA) was used to determine the mean significant differences in the pre-test and the post-test scores.

The paired sample t-test signed ranked test was applied to determine the mean and median significant differences between the pre-test and the post-test scores. P-value < 0.05 was considered statistically significant.

The crosstabs and the Chi-square test were used to examine the relationships between different variables in the data set. The crosstabs analysis involved constructing a table to display the frequency distribution of the data. The Chi-square test was employed to determine whether there was a significant association between the variables.

4 Results

The total number of participants in the summer short course was 574 in this study. The general characteristics and training variables were the descriptive statistics compiled.

4.1 Descriptive Statistics

Categorical variables, such as identity, curriculum, writing stage, papers published and difficulties faced, were presented in frequencies and percentages. In Table 1, the participants in the study comprised both master and doctoral students, with master students accounting for 69.81% in the experimental group and 72.22% in the control one. The doctoral students comprise 30.19% in the experimental group and 27.78% in the control.

Table 1. General characteristics of the study participants ($n = 574$)

Characteristics	Category	Experiment ($n = 520$)		Control ($n = 54$)	
		Participants	Percentage (%)	Participants	Percentage (%)
Identity	Master's students	363	69.81	39	72.22
	PhD students	157	30.19	15	27.78
Curriculum	No	285	54.81	39	72.22
	Yes	235	45.19	15	27.78
Writing stage	Plan	346	66.54	36	66.67
	Writing or publishing	174	33.46	18	33.33
Published papers	None	327	62.88	33	61.11

(*continued*)

Table 1. (*continued*)

Characteristics	Category	Experiment ($n = 520$)		Control ($n = 54$)	
		Participants	Percentage (%)	Participants	Percentage (%)
	At least 1 papers	193	37.12	21	38.89
Difficulties	Methods	429	82.50	44	81.48
	Language barriers	369	70.96	38	70.37
	Lack of guidance	120	23.08	10	18.52
	Insufficient time	67	12.88	10	18.52
	No companion or encouragement	66	12.69	10	18.52

The academic writing curriculum is usually offered by universities as an elective course for master and doctoral degree students. The curriculum contents include the writing structure, style and tone, etc., and may vary to some extent from institution to institution. Some participants have elected such writing relevant curriculum before the training program (45.19% in the experimental group, and 27.81% in the control), while others have never selected one yet (54.81% in the experimental group, and 72.22% in the control).

Summer holidays are also a good time for the master and PhD students to focus on preparing their research papers, whether they are in the planning, writing, or publishing stage. It is also a chance to review and identify potential target journals for publication. The survey shows that some participants are in the planning stage (66.54% in the experimental group, and 66.67% in the control), and others are in the writing or publishing stage (33.46% in the experimental group, and 33.33% in the control).

Part of the participants who have published at least one research article before the training course have writing experiences (37.12% in the experimental group, and 38.89% in the control), others have not (62.88% in the experimental group, and 61.11% in the control).

But all the participants are found to still face many difficulties in academic writing. The main reasons, according to a multi-choice question, include English language barriers, lack of guidance, insufficient time, and no companion or encouragement.

4.2 Independent Sample T-test

Does the short writing training program make a difference on the participants in the experimental group on improving their writing literacy?

In Table 2, the continuous variables of pre-test and post-test scores of the training were expressed as mean $\pm$ SD. The results show that the pre-test mean score of "writing awareness" of the participants is (13.59 ± 3.09), whereas the post-test mean score 15.38 ± 3.64. The pre-test mean score for correct responses for all the participants in the basic writing knowledge section is 11.61 ± 2.46, and the post-test mean score 12.82 ± 2.87. Regarding the scores of the participants in the "advanced writing knowledge" section, the mean score is 13.74 ± 3.02 in the pre-test while the post-test mean score 15.4 ± 3.55. Moreover, in the "writing tool application skills" section, the pre-test mean score is 8.68 ± 1.94 and the post-test mean score 9.73 ± 2.26. While the pre- and post-test in the control group are gentle and stable. In statistics, skewness and kurtosis are two ways to measure the shape of a distribution. Thus, there were higher scores in the pre-test than in the post-test. These differences were shown to be significant based on the results of an independent samples t-test.

Table 2. Difference of writing literacy by the training through the independent sample t-test/analysis ($n = 574$)

	Experiment group($n = 520$)				Control group($n = 54$)			
	Mean ± S.D	Median (Min.-Max.)	t	p	Mean ± S.D	Median (Min.-Max.)	Skewness	Kurtosis
Pre_Awareness	13.59 ± 3.09	14(5–24)	1.246	0.213	13.02 ± 4.25	14(5–22)	−0.801	−0.005
Post_Awareness	15.38 ± 3.64	15(5–25)	4.6571	< 0.01	12.91 ± 4.32	14(5–23)	−0.61	0.002
Pre_Knowledge A	11.61 ± 2.46	11(4–19)	2.863	0.004	10.56 ± 3.5	11(4–18)	−0.745	−0.01
Post_Knowledge A	12.82 ± 2.87	13(4–20)	5.374	< 0.01	10.56 ± 3.54	12(4–18)	−0.706	−0.118
Pre_Knowledge B	13.74 ± 3.02	14(5–24)	1.755	0.08	12.94 ± 4.262	14(5–22)	−0.712	−0.025
Post_Knowledge B	15.4 ± 3.55	15(5–25)	4.504	< 0.01	13.06 ± 4.37	15(5–23)	−0.689	−0.043
Pre_Skills	8.68 ± 1.94	9(3–14)	2.96	0.003	7.83 ± 2.6	9(3–12)	−0.769	−0.231
Post_Skills	9.73 ± 2.26	10(3–15)	5.577	< 0.01	7.89 ± 2.68	9(3–13)	−0.675	−0.263

Note Awareness = writing awareness, Knowledge A = basic writing knowledge, Knowledge B = advanced writing knowledge, Skills = writing tool application skills, Min. = Minimum, Max. = Maximum

4.3 Paired Sample T-test

The paired sample t-test was used to examine whether the training intervention has made a difference on the experimental group. The results obtained are displayed in Table 3. The data reveal that there is a significant difference between the pre- and the post-test in the writing literacy in the experimental group ($p < 0.01$). The training is obviously effective in improving the writing literacy of the participants with the interventions.

Table 3. Effects of writing literacy by the short-term summer training course by the paired sample t-test/analysis ($n = 574$)

					95% C.I			
		M.D	*S.D*	*S.E*	Lower	Upper	*t*	*p*
Experiment	Pre_Post_Awareness	−1.787	3.612	0.158	−2.098	−1.475	−11.28	< 0.01
	Pre_Post_Knowledge A	−1.206	2.903	0.127	−1.456	−0.956	−9.472	< 0.01
	Pre_Post_Knowledge B	−1.66	3.612	0.158	−1.971	−1.348	−10.476	< 0.01
	Pre_Post_Skills	−1.04	2.203	0.097	−1.23	−0.851	−10.771	< 0.01
Control	Pre_Post_Awareness	0.111	0.769	0.105	−0.099	0.321	1.062	0.293
	Pre_Post_Knowledge A	0	0.476	0.065	−0.13	0.13	0	1
	Pre_Post_Knowledge B	−0.111	0.718	0.098	−0.307	0.085	−1.137	0.261
	Pre_Post_Skills	−0.056	0.359	0.049	−0.154	0.042	−1.137	0.261

Note * $p < 0.05$ ** $p < 0.01$

4.4 Crosstabs Chi-Square Test

As mentioned above, the included participants in this study are either master or and doctoral students and some have taken elective writing curriculum before, or even have publishing experiences. The crosstabs analysis was observed with the post-test results in terms of study stage, previous writing curriculum taken or not, and with publishing experience or not.

This study also tries to determine whether the intervention has different impacts on those who have taken an elective writing curriculum versus those who have not, as well as on those who have published academic papers versus those who have not yet. The results show the effectiveness of the training intervention in improving the research article writing abilities to the different sub-group participants.

Table 4 shows the results on how the training intervention have impacted on the participants in the experimental group at different study stages, i.e. either reading a master or a doctoral degree. In general, doctoral students have better writing skills than master degree students. In this study, after the training intervention, the writing awareness of the participants shows significant difference ($p = 0.048$).

Table 4. Post-test score and study stages crosstabs analysis

			n	≤ 12	% within stage (%)	% of total (%)	> 12	% within stage (%)	% of total (%)	χ^2	p
Post_Awareness	Experiment	Master	363	71	19.56	13.65	292	80.44	56.15	30.324	0.048*
		PhD	157	17	10.83	3.27	140	89.17	26.92		
		Total	520	88	16.92	16.92	432	83.08	83.08		
	Control	Master	39	15	38.46	27.78	24	61.54	44.44	10.491	0.312
		PhD	15	2	13.33	3.70	13	86.67	24.07		
		Total	54	17	31.48	31.48	37	68.52	68.52		
Post_Knowledge A	Experiment	Master	363	200	55.10	38.46	163	44.90	31.35	25.224	0.066
		PhD	157	66	42.04	12.69	91	57.96	17.50		
		Total	520	266	51.15	51.15	254	48.85	48.85		
	Control	Master	39	36	92.31	66.67	3	7.69	5.56	18.04	0.054
		PhD	15	10	66.67	18.52	5	33.33	9.26		
		Total	54	46	85.19	85.19	8	14.81	14.81		
Post_Knowledge B	Experiment	Master	363	63	17.36	12.12	300	82.64	57.69	15.267	0.761
		PhD	157	19	12.10	3.65	138	87.90	26.54		
		Total	520	82	15.77	15.77	438	84.23	84.23		
	Control	Master	39	14	35.90	25.93	25	64.10	46.30	22.193	0.023
		PhD	15	2	13.33	3.70	13	86.67	24.07		
		Total	54	16	29.63	29.63	38	70.37	70.37		
Post_Skills	Experiment	Master	363	338	93.11	65.00	25	6.89	4.81	19.085	0.087
		PhD	157	143	91.08	27.50	14	8.92	2.69		
		Total	520	481	92.50	92.50	39	7.50	7.50		
	Control	Master	39	39	100.00	72.22	0	0.00	0.00	19.308	0.023
		Yes	15	15	100.00	27.78	0	0.00	0.00		
		Total	54	53	98.15	98.15	1	1.85	1.85		

Note * $p < 0.05$ ** $p < 0.01$

Table 5 shows the results on how the training intervention have impacted on the participants in the experimental group who have taken a writing curriculum in their first or second semester before this training program over those who have not any in different study stages. In general, doctoral students have better writing skills than master students. In this study, after the training intervention, the writing awareness and the basic writing knowledge of the participants shows significant difference, respectively ($p = 0.019$, $p = 0.012$).

Table 5. Post-test score and curriculum experience crosstabs analysis

			n	Score ≤ *12*	% within cur (%)	% of total (%)	Score > 12	% within cur (%)	% of total (%)	χ^2	*p*
Post_Awareness	Experiment	No	285	56	19.65	10.77	229	80.35	44.04	33.863	0.019*
		Yes	235	32	13.62	6.15	203	86.38	39.04		
		Total	520	88	16.92	16.92	432	83.08	83.08		
	Control	No	39	14	35.90	25.93	25	64.10	46.30	8.612	0.474
		Yes	15	3	20.00	5.56	12	80.00	22.22		
		Total	54	17	31.48	31.48	37	68.52	68.52		
Post_Knowledge A	Experiment	No	285	156	54.74	30.00	129	45.26	24.81	31.511	0.012*
		Yes	235	110	46.81	21.15	125	53.19	24.04		
		Total	520	266	51.15	51.15	254	48.85	48.85		
	Control	No	39	34	87.18	62.96	5	12.82	9.26	7.793	0.649
		Yes	15	12	80.00	22.22	3	20.00	5.56		
		Total	54	46	85.19	85.19	8	14.81	14.81		
Post_Knowledge B	Experiment	No	285	51	17.89	9.81	234	82.11	45.00	26.713	0.144
		Yes	235	31	13.19	5.96	204	86.81	39.23		
		Total	520	82	15.77	15.77	438	84.23	84.23		
	Control	No	39	13	33.33	24.07	26	66.67	48.15	7.960	0.717
		Yes	15	3	20.00	5.56	12	80.00	22.22		
		Total	54	16	29.63	29.63	38	70.37	70.37		
Post_Skills	Experiment	No	285	264	92.63	50.77	21	7.37	4.04	15.524	0.214
		Yes	235	217	92.34	41.73	18	7.66	3.46		
		Total	520	481	92.50	92.50	39	7.50	7.50		
	Control	No	39	38	97.44	70.37	1	2.56	1.85	5.504	0.788
		Yes	15	15	100.00	27.78	0	0.00	0.00		
		Total	54	53	98.15	98.15	1	1.85	1.85		

Note * $p < 0.05$ ** $p < 0.01$

Table 6 shows the results on how the training intervention have impacted on the participants in the experimental group whether they have published papers or not before this training program. Table 6 shows that the participants who have published English research articles have better basic writing knowledge and have trained writing awareness to some extent. Through this training intervention, their advanced writing knowledge have got improved significantly ($p = 0.006$).

Table 6. Post-test score and published experience crosstabs analysis

			n	≤ 12	% within papers (%)	% of total (%)	> 12	% within papers (%)	% of total (%)	χ^2	p
Post_Awareness	Experiment	Have	193	19	9.84	3.65	174	90.16	33.46	35.924	0.11
		None	327	69	21.10	13.27	258	78.90	49.62		

(continued)

Table 6. (*continued*)

			n	≤ 12	% within papers (%)	% of total (%)	> 12	% within papers (%)	% of total (%)	χ^2	p
		Total	520	88	16.92	16.92	432	83.08	83.08		
	Control	Have	21	5	23.81	9.26	16	76.19	29.63	3.784	0.925
		None	33	12	36.36	22.22	21	63.64	38.89		
		Total	54	17	31.48	31.48	37	68.52	68.52		
Post_Knowledge A	Experiment	Have	193	88	45.60	16.92	105	54.40	20.19	22.824	0.119
		None	327	178	54.43	34.23	149	45.57	28.65		
		Total	520	266	51.15	51.15	254	48.85	48.85		
	Control	Have	21	16	76.19	29.63	5	23.81	9.26	8.833	0.548
		None	33	30	90.91	55.56	3	9.09	5.56		
		Total	54	46	85.19	85.19	8	14.81	14.81		
Post_Knowledge B	Experiment	Have	193	25	12.95	4.81	168	87.05	32.31	39.605	0.006**
		None	327	57	17.43	10.96	270	82.57	51.92		
		Total	520	82	15.77	15.77	438	84.23	84.23		
	Control	Have	21	5	23.81	9.26	16	76.19	29.63	15.586	0.157
		None	33	11	33.33	20.37	22	66.67	40.74		
		Total	54	16	29.63	29.63	38	70.37	70.37		
Post_Skills	Experiment	Have	193	175	90.67	33.65	18	9.33	3.46	18.747	0.095
		None	327	306	93.58	58.85	21	6.42	4.04		
		Total	520	481	92.50	92.50	39	7.50	7.50		
	Control	Have	21	20	95.24	37.04	1	4.76	1.85	11.611	0.236
		None	33	33	100.00	61.11	0	0.00	0.00		
		Total	54	53	98.15	98.15	1	1.85	1.85		

Note * $p < 0.005$ ** $p < 0.001$

However, there were no difference observed in terms of the writing tool application skills.

5 Conclusion

We studied how academic writing training can improve the writing literacy to early career researchers, graduate and doctoral students in special, whose English is a second language through a 3.5-day short-term online summer training program conducted in the NSL. The results show that the writing awareness, the basic writing knowledge, the advanced knowledge, and the writing tool application skills have all been improved to some extent in the study. Based on the findings of this study, we strongly recommend it and plan to conduct the similarly structured summer academic writing training program every year in the NSL, which is also valuable for other academic libraries to explore author education activities. This pioneering study proves the effectiveness of the training on promoting the academic writing literacy development for non-native English-speaking researchers.

References

ACRL.: Framework for information literacy for higher education [EB/OL] (2023). http://www.ala.org/acrl/sites/ala.org.acr/files/content/standards/framework-chinese.pdf

Mullen, C.A.: The need for a curricular writing model for graduate students. J. Furth. High. Educ. **25**(1), 117–126 (2001)

Emma, M., Carole, J.T.: Single group, pre- and post-test research designs: some methodological concerns. Oxford Rev. Educ. **38**(5), 583–616 (2012)

Engaging Citizens and Communities: A Public History Approach for Libraries

Chiara De Vecchis(✉)

Italian Library Association (AIB—Associazione Italiana Biblioteche), Rome, Italy
chiara.devecchis@aib.it

Abstract. By highlighting the affinities between Public History and libraries as experts in audience engagement, the paper proposes activities and strategies, inspired by the Public History approach, that libraries may perform—also in cooperation with other GLAM communities—to promote historical consciousness and cultural heritage as a vehicle of knowledge and social inclusion, therefore enhancing the connection between libraries and democracy, so that it may become more effective.

Keywords: Cultural Heritage · GLAM · Public Engagement

1 Introduction

Libraries are—or we believe they should be—agents of connection. Can they also connect people with their history, making them perceive the past as living and operating in the present? If so, is this a goal that libraries are specifically meant to achieve? Can't we just rely on the idea that only libraries with historical collections, or dealing with historical disciplines in academic contexts, should be involved in how history is taught and transmitted? And, above all: does it really matter?

As professionals whose role deals with the shaping and sharing of knowledge, we could agree that every knowledge matters. We can't opt out: history concerns librarians, whose commitment is to act transformatively within nowadays' complex society; moreover, history concerns all of us, as citizens who take part in the community we're living in; finally, history concerns us as human beings connected to each other. Acting as promoters of historical knowledge among communities contributes to civic awareness, therefore becoming an effective vehicle for democracy, as professional literature assumes libraries (and librarians) to be. From this point of view, libraries may play a pivotal role in promoting a shared vision of people's roots—against simplifications and despite ideologies—as a driving force for social inclusion, helping overcome the barriers made by different representations of history. The cognitive bias hidden in sticking to self-representations of social groups is not only an issue for historians; it's also a pitfall that should be countered with lifelong education, a field that's a natural arena for libraries and the lifeblood for their dialogue with communities.

Assuming all this, the question is: how could libraries manage such a challenge? The purpose of this paper is to suggest a range of possible answers, by introducing a

A. Katsirikou (Ed.): QQML 2023, 68, pp. 49–58, 2026.
https://doi.org/10.1007/978-3-032-01264-7_4

new approach to some of the library activities through the lens of Public History and promoting it as a new way of engaging communities. To achieve this goal, we can start by briefly considering the context; then, we'll recall a few definitions from the literature on Public History; we'll investigate connections with the library field and propose some strategies inspired by worldwide experiences; finally, we'll discuss their outputs and try to look to the future.

2 Background

2.1 Society and Cultural Heritage

Libraries' commitment to engaging communities and to raising citizens' awareness has even increased in the last decades, partly by coping with social changes and new technologies, which may imply, in turn, opportunities and challenges.

A growing phenomenon that libraries should consider is that almost any kind of online communication (web, social networks) acts as an amplifier, giving voice to self-representations of people and social groups seeking visibility, especially within the communities whose stories have been untold, silenced, miscommunicated or just neglected by "official" history: for example, ethnic or linguistic minorities, LGBTQ + communities—and the list could continue. This phenomenon fosters a widespread need for history, not considered just as a chronological chain of facts, but as a search for roots, perceived in deep connection with people's lives and identities, so that the past may be "felt" as active in the present to build the future.

The trend is strengthened by the debate on cultural heritage—a global reflection drawing a revolution in cultural policies during the last decades. Just to recall a couple of milestones, the Unesco *Convention for the Safeguarding of the Intangible Cultural Heritage*, issued in 2003, pointed out the role of communities, stating that they «play an important role in the production, safeguarding, maintenance and re-creation of the intangible cultural heritage, thus helping to enrich cultural diversity and human creativity» [1]; a few years after, the *Convention on the Value of Cultural Heritage for Society* (signed in Faro, Portugal, in 2005 and in force since June 2011, but still being ratified by some countries) was based upon the idea that «rights relating to cultural heritage are inherent in the right to participate in cultural life, as defined in the Universal Declaration of Human Rights»; it also acknowledged «the role of cultural heritage in the construction of a peaceful and democratic society and in the processes of sustainable development and the promotion of cultural diversity» [2]. Since the Faro Convention quotes article 27 of the *Universal Declaration of Human Rights*, stating that «everyone has the right freely to participate in the cultural life of the community», let's also recall the text of article 22 in the same Declaration: «Everyone, as a member of society, [...] is entitled to realization [...] of the economic, social and cultural rights indispensable for his dignity and the free development of his personality» [3]. As a consequence, if dealing with cultural heritage must be considered among human rights, then the knowledge of the past, as an important part of our culture, stops being just an educational requirement and becomes a deeply personal matter for anyone.

Furthermore—when talking about global horizons, cultural heritage and sustainability—we can't forget the *2030 Agenda for Sustainable Development* [4]. Helping historical knowledge to be widespread into society is indeed connected with at least two of the Sustainable Development Goals promoted by the United Nations, namely the ones ensuring lifelong, quality education (Goal 4, which has been recently mentioned among the priorities to drive transitions) and promoting peaceful, inclusive, participatory societies, through actions to strengthen our institutions (Goal 16, target 16.7) [5]. Not surprisingly, the same goals are among those most likely to be met by the library action.

2.2 From the Academic Backdrop to the Communication of History

It should therefore be clear to what extent our visions of the past can impact on our present and on our future: such consciousness may be an essential part of the soil on which libraries build their social action as inclusive mediators of knowledge.

Unfortunately, it's not so easy. On one hand, academic communication—which is a fundamental pillar in the transmission of knowledge—is hardly able to meet such a widespread need for history, nor to elicit a response in terms of emotional participation: it is oriented on producing scientific papers and raising researchers and specialists, not on dealing with the general public or engaging wide audiences. It doesn't have among its priorities the use of a popular language and it's not aimed to convince people that competences in history may be useful for them. The distance between academic communication and the general public, widened by social media, has played a role in the growing—and more general—phenomenon of distrust towards science; people are often «interested in the past but not in the history as it was defined in academic textbooks» [6].

On the other hand, we cannot—nor do we want to—reduce the authority or ignore the results of scientific and historical research, in the firm belief that the scientific method is a fundamental bulwark against the drift of disinformation. We must be aware that anyone can claim to be an "expert" in something, but not everybody knows or is able to use the historical method, not everybody has the necessary skills and competences for working on historical sources, or for evaluating and interpreting new kinds of evidence such as digital and new media sources. Recent trends show that uncontrolled and scarcely aware circulation of historical information, combined with a growing disbelief in science, may lead to pseudoscience.

If the knowledge of the past is partial, affected by cognitive bias, not based on a solid background, or not paying attention to the difference between history and memory, between facts and opinions, it may indeed contribute to misinformation and to fake news: something that libraries are committed to countering through continuing education and information tools (the IFLA infographic on spotting fake news is just one of the most renowned examples) [7]. Moreover, a weak historical education, coupled with poor quality information, may leave the door open to intentional (ab)uses of historical memory, such as manipulating the interpretation of the past in order to meet political purposes or to foster consensus around a party. The quest for social inclusion through historical knowledge, if not based on solid sets of skills, may turn into the risk of new divisions.

We can consider all of this as the *humus*, the deep social need from which Public History was born as a response—and an antidote. But what exactly is Public History? This is still a concept not everybody is familiar with; how can we define it?

3 Around Public History

3.1 What Public History Is…

Public History (δημόσια ιστορία in Greek) [8] is a worldwide movement, yet—as any growing phenomenon—it still needs to fight against its trivialization or misunderstandings, and even against its uncertain self-determination. A review of the fast-growing literature on the topic—which goes beyond the intentions of this paper—would in fact highlight «the lack of an agreed upon definition», often replaced by the promotion of «a big-tent concept of the field» [9]. This is why, for the purposes of this paper, we'll stick to an authoritative and widely accepted interpretation, the one from the International Federation for Public History. Its goal is to devise a broad concept of Public History at global level, encompassing a variety of context-sensitive declinations that occur and take hold in different countries or areas of the world. IFPH states that «Public History international is a field in the historical sciences made up of professionals who undertake historical work in a variety of public and private settings for different kinds of audiences worldwide. The settings in which they work may include international and transnational organizations, governments at all levels, as well as local, regional, or national non-profit, corporate, cultural and educational institutions» [10].

Cooperation is then a highly recommendable feature, also pointed out by the US National Council on Public History: «Unlike many historians in the academy, public historians routinely engage in collaborative work, with community members, stakeholders, and professional colleagues, and some contend that collaboration is a fundamental and defining characteristic of what public historians do» [11]. This may of course include—according to the *Italian Public History Manifesto*—cooperating with librarians: «Public Historians work in cultural institutions, museums, archives, libraries, media, in the cultural and the tourism industry, in schools, and are engaged in cultural volunteering and social promotion and in all fields where the knowledge of the past is required to work with different audiences» [12].

The purpose of cooperation is to elaborate historical knowledge *for* the public and *with* the public; the context is the whole society, not only universities. Far from ignoring the importance of academic history—and learning from it, rather—a peculiarity of Public History relies in finding new ways of engaging people in understanding the past as a living part of the present, thus encouraging active citizenship.

To do so, new languages are required, transcending disciplinary divisions and removing barriers between the academic approach and the professional practice. Some of the actions performed outside—but not in competition with—academic settings include history festivals, public debates, documentaries, practices of re-enactment or theatrical adaptations of historical events, bibliographic and documentary exhibitions, gaming on historical topics, commemorating anniversaries, building historical parks, developing projects about the meaning of monuments or toponymy and the memories they may convey, but also sharing and spreading information and competences on the evaluation, interpretation and management of historical sources, such as those collected by archives and libraries.

3.2 … And What It's Not

Anyway, since the purpose of this paper is not to define Public History but to connect its action with library action, therefore being aware of what Public History *is not* may also help keeping away from misunderstandings and bad practices.

Public History is sometimes confused with scientific dissemination or even with the popularization of historical concepts; yet, it has a slightly different approach. It is not only about narrating history in an easy way; it focuses on history production. It's not a bibliographic mix of secondary sources merged in appealing outputs; it highlights key sources and teaches about them. It's not a one-way communication towards passive spectators; it aims to engage audiences so that people may share authority with historians [13] and be able to bring a new awareness into their working practices—we might say, into their lives. A successful action, in the field of Public History, should not only end in the acquisition of historical knowledge (an important goal indeed, in traditional learning); it should also generate, widen and spread new competences in research methods.

Last but not least, Public History aims to match and fulfil people's need—even when unconscious—for history: it is sometimes described as an applied history and promotes history as a common good, which is necessary for collective memory and for democratic participation. The public historian is a mediator, somewhat like the conductor of an orchestra where citizens and professionals (including librarians) interact and cooperate to build stronger institutions and societies.

It's a challenging goal, indeed. When the going gets tough, will libraries get going?

4 Strategies of Cooperation and Engagement

4.1 The Role of Libraries

We've just reviewed some goals in terms of engaging audiences, sharing skills, teaching methods, facilitating knowledge, activating connections, undergoing democratic processes: is it enough to consider Public History and libraries as cooperating allies? Wesley Johnson seemed not to doubt it when, almost fifty years ago, he pointed out the areas of interest for Public History, finding a common denominator with libraries in their role of mediators [14]. More recently, Jeff Manuel emphasized the «natural affinity» between Public History and public libraries, highlighting the value of libraries «as forums for collaboration between university-based public history programs and local communities» [15].

Actually, libraries act as hubs and managers of historical sources; they are skilled in evaluating and interpreting documents; they are widespread in different places and communities as expert communicators and facilitators of information literacy; they are user-centered; through their services—but also through the study and enhancement of their collections—libraries can be active places of the dialectic between politics and history, local and global, institutional and personal experience, past and present, thus educating the society to history and citizenship [16]; in one word, they may become «agents of public history» [17].

It shouldn't be surprising, then, that libraries are suitable to enhance their action by considering the Public History approach, in cooperation with historians, as an opportunity to advocate for themselves among the scientific community, but also as a less usual—though effective—way to reach their institutional purposes. This may be achieved also by cooperating with archives and museums, which are main partners in the preservation and promotion of cultural heritage as an important part of our past: GLAM communities, indeed, are often involved in the activities of the associations of public historians.

These intersections are sometimes—yet, still too seldom—underlined by the literature on Public History, which often points out that libraries are spaces where historically informed interactions take place. This doesn't just mean hosting Public History projects or events in the library but—and much more—devising new strategies according to the point of view of the public historian, in the horizon of a new model of library, matching online collaboration and public engagement in a user-centered perspective [18].

4.2 From Traditional to New Practices in the Digital Environment

A quick overview of good practices may start by considering a wide range of frequent activities in libraries, from the most traditional such as bibliographic exhibitions, up to information literacy classes or even to practices of media communication, gaming, re-enactment, which may be interpreted by libraries with a Public History twist, whenever such activities are meant to engage well-targeted audiences, right from the design stage. The QQML23 Conference itself actually hosted many papers promoting experiences which were not born to be Public History practices, yet are suitable to inspire them, as long as they could be interpreted or implemented in the key of Public History.

Recent evolutions in digital issues seem to accelerate the process of defining innovative methods for library action in this area; Public History itself is increasingly referred to as Digital Public History [19]. In particular, the development of Digital Humanities in academic contexts, together with the huge growth of user-generated content (which makes all of us not only consumers but also producers—or "prosumers"—of information), play an important role both for libraries and Public History, defining a wide, articulated area of interconnections, and suggesting some strategies for library action in this field. Practices may include—but are not limited to—bringing together a range of digital resources related to a topic or to a historical period, with federated search tools (e.g. Connected Histories, a project on British history sources, 1500–1900) [20]; collecting sources and memories (e.g. Community archives and identities project) [21]; building rapid-response archives to share personal stories and experiences related to historical, widely-spread phenomena (Covid-19 was, unfortunately, a suitable occasion for such activity [22]; developing practices which involve crowdsourcing; and co-curating.

4.3 Crowdsourcing and Co-curating

These last two practices seem the most relevant to take into consideration when it comes to showing the variety of possible actions that libraries can carry out or be involved in as partners or promoters of Public History projects. Crowdsourcing is considered one of the most interesting and original ways to make people interact with library collections—and with history [23]. It can be generally interpreted as the collecting and sharing of relevant documents (texts, images, data: e.g., diaries, photographs, digital memories), from and among users, to meet specific goals, e.g. to build community archives (usually to be shared on the internet) under the guidance of the public historian and/or librarian; nevertheless, in its more interesting digital applications [24], crowdsourcing may mean asking for users' help in managing and processing data collections: for instance, crowdsourcing transcriptions of documents in ongoing digitization projects, as in the New York Public Library project *What's on the Menu?* [25].

Crowdsourcing is often combined with co-curation, a widespread trend that highlights the growing importance of the curatorial role—in a broad sense—in today's culture [26]. Co-curating may mean, in the library field, establishing a cooperation with the public in designing cultural initiatives; inviting users to bring their own expertise and experience in interacting with collections; promoting actions that sometimes border on citizen science experiments, such as the *Transcribathon* promoted by Europeana. This project—whose slogan wishes «to enrich digital cultural heritage using the power of the crowd»—involves a competition in the processing of selected documents (images, postcards, letters, diaries) collected by a number of cultural heritage institutions: people are encouraged «to work with historical material and become an integral part of their digitization process» by «adding transcriptions, tags and georeferences»—or even translations and hints to trace the date of the documents—whose potential for further research is thus exploited [27]. As a result, hopefully, citizens working on historical documents may feel connected with the 'stories' and memories they convey; as a collateral effect, they come in touch with the library as a collector of sources but also—and even more—as an inclusive educational institution.

4.4 Challenges and Opportunities

Of course, such initiatives may face risks, mainly connected to the quality of the data collected and to impact assessment. How are we going to check the data we have obtained from crowdsourcing and the overall quality of the final output? How can we measure, in terms of knowledge acquisition, the impact of participating in similar activities, unless we provide opportunities for in-depth study or more formal learning? How can we learn more about the cultural background of the participants and be sure of what they have gained from the project? Will participants be able to consider documents not just as singularities to be tagged, but as parts of a complex mosaic of interrelated sources? Will they be able to build meaningful connections with documents other than the ones they've been working on?

Some of these pitfalls may be minimized by professionals, by taking care of the whole process, monitoring it constantly, inserting feedback mechanisms and, most of all, by remembering the distance between crowdsourcing and—something that libraries should never do—disintermediation: mediation between people and documents is what libraries are specifically meant to perform. On the other hand, a common feature and also a consequence of these kinds of practices is an expanded vision of library users: the "public" is interpreted as an audience—large but not indefinite nor random—both inside and outside the library, engaged in dynamics of trust and participation to the life of the library itself.

A final consideration may be introduced by briefly recalling the experience of the Italian Library Association (AIB), which sits among the founders of the Italian Association of Public History (AIPH). One of the main commitments for the library association in this field was to define a paradigm of library action for Public History: to do so, during the latest years, best practices of Public History in libraries have been collected throughout the Country; surveys and debates have been promoted among librarians; the outputs have been presented and discussed in every AIPH conference of Public History, held each year in a different Italian town (Ravenna, Pisa, S. Maria Capua Vetere and Caserta, Mestre and Venice, Rome) [28]. It was soon found out that Public History is of cross-cutting interest to any typology of library—public, academic, school, special libraries—as a driving force for social inclusion, a tool for interconnecting local communities, a strategy to advocate for researchers and scholars, a way to educate and help build strong identities, and an opportunity for collecting and connecting different kinds of sources (e.g. archival documents, personal collections). Each library, in fact, addresses people, and engaging people is a common denominator both for Public History and for the 'new' librarianship, as long as «the mission of librarians is to improve society through facilitating knowledge creation in their communities» [29].

5 Conclusions?

In a rapidly evolving context, "conclusions" should not really end the debate; they rather suggest a new starting point.

If we were asked to draw in a summary figure the many intersections we have considered, we could use a cyclic image: in a digital environment, Public History encourages an interdisciplinary dialogue which involves historians, librarians and other GLAM professionals, who in turn engage communities to foster social inclusion and democratic participation, also producing other digital outputs, and so on, in an overall logic of social welfare.

A more suggestive image comes from literature on Public History. A few years ago, Thomas Cauvin, former president of IFPH, imagined Public History as a tree, symbol of a systemic process where libraries—collecting and providing documents—deal with the roots, by managing sources to be then interpreted (the trunk of the tree) and communicated (the branches: books and media), so to be used (the leaves) [30].

If we are able, as a professional community, to communicate our role in all the mentioned intersections, then it will become clear that libraries are not only "roots" for Public History, as managers of sources; libraries act, indeed, wherever documents are (in schools, in media, in social networks), up to the branches of this tree (communicating history) and to its leaves (uses of history). Libraries deal with education, identity, fun, empowerment, politics, social justice—and the list goes on. We hope that, in the future, the same image of the tree can make libraries appear as if they flow through the sap of the tree itself, because libraries are everywhere people (and their history, and their need for knowledge) are—or, at least, we believe they should be.

References[1]

1. Unesco: Convention for the safeguarding of the intangible cultural heritage (2003), [Preamble]. https://ich.unesco.org/en/convention
2. Council of Europe: Convention on the value of cultural heritage for society (2005), art. 1.a. https://rm.coe.int/1680083746
3. United Nations: Universal declaration of human rights (1948). https://www.un.org/en/about-us/universal-declaration-of-human-rights
4. United Nations: Transforming our world: the 2030 agenda for sustainable development (2015). https://undocs.org/en/A/RES/70/1
5. United Nations: The sustainable development goals report. Special edition, pp. 50–51 (2023). https://unstats.un.org/sdgs/report/2023/
6. Apostopoulos, P.: What is the public of public history? Between the public sphere and public agency. Magazèn, Int. J. Digit. Public Hum. **2**(2), 311–327 (2021). https://doi.org/10.30687/mag/2724-3923/2021/04/006
7. International Federation of Library Associations and Institutions—IFLA: How to spot fake news (2017). https://repository.ifla.org/handle/123456789/167
8. Peponas, M.: Public history in Greece. Bridging, February 11 (2019). https://ifph.hypotheses.org/2748
9. Conard, R.: Still grappling with the definition question. Public Hist. **1**(40), 115–119 (2018). https://doi.org/10.1525/tph.2018.40.1.115
10. International Federation for Public History—IFPH Homepage. https://ifph.hypotheses.org/
11. [US] National Council on Public History—NCPH: About the field. https://ncph.org/what-is-public-history/about-the-field/
12. Associazione Italiana di Public History—AIPH: The Italian public history manifesto, December 13 (2018). https://doi.org/10.58079/awly
13. Frisch, M.: A shared authority. Essays on the craft and meaning of oral and public history. State University of New York, Albany (1990)
14. Johnson, G.W.: Editor's preface. Public Hist. **1**(1), 4–10 (1978). https://doi.org/10.2307/3377660
15. Manuel, J.: Public history and public libraries. A natural affinity. History@Work. The NCPH blog, April 28 (2015). https://ncph.org/history-at-work/public-history-and-public-libraries/
16. Sabba. F.: The library heritage's enhancement between public engagement and public history. AIB Studi, Rivista di Biblioteconomia e Scienze dell'Informazione **1**(60), 9–18 (2020). https://doi.org/10.2426/aibstudi-12025
17. Arboleda, J.I.: Libraries as agents of public history: the case of the Luis Angel Arango library in Colombia, [paper presented at] 3rd IFPH Conference (2016)
18. Väljataga, M.: Digital public history in libraries. In: Noiret, S., Tebeau, M., Zaagsma, G. (eds.) Handbook of Digital Public History, pp. 185–198. De Gruyter Oldenbourg, Berlin-Boston (2022). https://doi.org/10.1515/9783110430295-016
19. Salvatori, E.: Digital public history inside and outside the box. Magazèn, Int. J. Digit. Public Hum. **2**(1), 203–222 (2020). https://doi.org/10.30687/mag/2724-3923/2020/02/003/
20. University of Hertfordshire, University of London, University of Sheffield, (2011–2018), University of Sheffield (2019): Connectedhistories.org, version 1.0. https://www.connectedhistories.org/

[1] Further references may be easily found in ELPHi—Electronic Library of Public History: an online, open access, international, digital library on Public History, hosted by the academic repository of Salerno University in Italy: http://elea.unisa.it/handle/10556/4878. The web resources mentioned in the paper were last accessed 2024/02/28.

21. University College London: Community Archives and Identities: Documenting and Sustaining Community Heritage (2008). https://www.ucl.ac.uk/information-studies/community-archives-and-identities-documenting-and-sustaining-community-heritage
22. Arizona State University School of Historical, Philosophical and Religious Studies: Journal of the plague year. An archive of Covid-19. Web Archive (2020). https://covid-19archive.org/s/archive/page/Share
23. Holley, R.: Crowdsourcing: how and why should libraries do It? DLib Magazine, March-April (2010), [1–21]. http://eprints.rclis.org/14360/
24. Noiret, S.: Crowdsourcing and user-generated content, the raison d'être of digital public history? In: Noiret, S., Tebeau, M., Zaagsma, G. (eds.) Handbook of Digital Public History, pp. 35–48. De Gruyter Oldenbourg, Berlin-Boston (2022). https://doi.org/10.1515/9783110430295-003
25. New York Public Library—NYPL Labs: What's on the Menu? Homepage. https://menus.nypl.org/
26. Tebeau, M.: Curation. Toward a new ethic of digital public history. In: Noiret, S., Tebeau, M., Zaagsma, G. (eds.) Handbook of Digital Public History, pp. 277–290. De Gruyter Oldenbourg, Berlin-Boston (2022). https://doi.org/10.1515/9783110430295-023; O'Neill, P.: The curatorial turn: from practice to discourse. In: Rugg, J., Sedgwick, M. (eds.) Issues in Curating Contemporary Art and Performance, pp. 13–28, The University of Chicago Press, Chicago (2007). https://mnuai.ro/wp-content/uploads/2023/05/null-THE-CURATORIAL-TURN-FROM-PRACTICE-TO.pdf
27. Europeana: Transcribathon. Homepage. https://transcribathon.eu/
28. De Vecchis, C.: Libraries and public history: intersections, opportunities, challenges. Biblioteche Oggi Trends. **1**(7), 32–42 (2021). https://doi.org/10.3302/2421-3810-202101-032-1
29. Lankes, D.: A new librarianship for a new age, 57th AIB CONFERENCE. Rome, November 18 (2011). https://davidlankes.org/rdlankes/Presentations/2011/RomeEN.html
30. Cauvin, T.: New field, old practices: promises and challenges of public history, Magazèn, Int. J. Digit. Public Hum. **1**(2), 13–44 (2021). https://doi.org/10.30687/mag/2724-3923/2021/03/001

Digital Informative Education: Tackling Disinformation Through New Models of Communication of Libraries

Anthi Katsirikou[1], Iphigenia Georgakopoulou[2(✉)], Anna Theofilatou[2], and Eleana Pathiakaki[3]

[1] University of Piraeus Library, Athens, Greece
[2] National Library of Greece, Athens, Greece
georgakopoulou@nlg.gr
[3] Athens School of Fine Arts Library, Athens, Greece

Abstract. The life cycle of information begins with an information need with a special role in libraries. Knowledge of informational sources and especially of electronic sources are recognized as vital factors at all levels of research with higher literacy needs than any other time. At the same time, it is considered as given for every organization to have and develop its own digital presence via the Internet. Thus, through digital marketing, an organization or agency, with appropriate management, can effectively communicate with its environment and audience by utilizing digital tools and social networks, and through the balance of these and traditional communication channels. The challenge of the above is the creation of a common language of communication in the new online world of information, based on International Information Education Standards. In this educational context, the participation of collaborative networks and ongoing information literacy programs based on International Standards shape cultural values and abilities in users to produce independent, directed scientific research. The phenomenon of fake news and misinformation is a risk/danger. Scholars De Paor and Heravi (Information literacy and fake news: how the field of librarianship can help combat the epidemic of fake news, 2020), studied how the field of librarianship can help combat the phenomenon of fake news and disinformation. The IPS model of information behavior by Brand-Gruwel, Wopereis, and Walraven (Research Information—a descriptive model of information problem solving while using internet, 2009) describes informational problem reception (IPS) in which the Internet is involved in locating information. The Big6 Skills model of information education, which is a top-down approach focusing on the general in the process of total problem solving before focusing on the details and works in a circular motion, including project definition, information search strategies, identification and information access, use, synthesis and evaluation. Each of these include the stages for the correct use of the information. The Big Six Skills method can be a dynamic tool for the development of information literacy skills with the basic one being that of the correct use of information with basic steps being contact with information sources (hearing, sight, reading) and the correct identification of similar sources. A combination of this type can improve the digital gap as it can and does reach a diverse audience with different skills, with feedback and problem solving capabilities.

A. Katsirikou (Ed.): QQML 2023, 68, pp. 59–64, 2026.
https://doi.org/10.1007/978-3-032-01264-7_5

Keywords: Disinformation · Misinformation · Fake news

1 Introduction

People's need for information is mostly developed by the stimuli they receive. Stimuli from many areas that make up their daily lives such as social, economic and political issues. Therefore, each time a user searches for a topic, it should be understood that this topic can be found in different sources, and each time it is formulated through a different perspective (Korobili & Togia, 2016).

As society moves further into the digital age, the nature of information consumption, assimilation and circulation has changed drastically. The overconsumption of information fueled by the Internet has created a so-called "post-truth" society in which people consume information that confirms their pre-existing beliefs and ideologies rather than trying to identify the truth. As the online environment becomes more complex and difficult to navigate, doubt and uncertainty continue to increase with people more likely to value known facts that are more familiar to them than reliable ones (Gallagher, 2016).

1.1 Fake News

As part of the change in people's information behavior, the emergence and popularity of social media has complicated the relationship between people and information further facilitating the spread of fake news. Fake news creators have taken advantage of this social trend with an ultimate purpose the intentional flow of fake news through crowdsourcing tools with a strong element of bias not to the extent of creating personal conclusions but propaganda.

Information literacy, information agencies, family, school, science will face the challenges of scattered, unfiltered and stormy information in terms of educating young people, developing their ability to filter and control information in terms of accuracy and reliability of sources until she comes of age.

There have been several theoretical models being proposed in describing the necessary competencies expected from an information literate individual (Anunobi & Udem, 2014). The Big 6 information literacy skill model propounded by Eisenberg and Berkowitz (Berkowitz, 1990) is described as a systematic information behavior (Bruce, 2004) which explains the processes undergo by an individual to accomplish the competencies in becoming information literate. The model is not linear and circular movement is reinforced in the process of solving informational problems and this can be repeated many times and not in the same order if this is required for the best performance of the result.

2 EU—Shaping Europe's Digital Future (Commission, 2022)

The Commission is tackling the spread of online disinformation and misinformation to ensure the protection of European values and democratic systems. Disinformation is false or misleading content that is spread with an intention to deceive or secure economic or political gain, and which may cause public harm. Misinformation is false or misleading content shared without harmful intent though the effects can be still harmful. Disinformation is an evolving challenge, with high potential to negatively influence democratic processes and societal debates. Its increasingly adverse effects on society across the European Union call for a coordinated, joint and sustainable approach to comprehensively counter it. Psychological research looks at individual differences in demographic, personality, and other traits of those who are more likely to believe misinformation and conspiracy theories, with the ultimate goal of characterizing the underlying processes that lead people to accept such claims. The findings outline some individual differences psychologists have identified, but they should not be used to generalize across groups regarding belief in misinformation.

- 80% of Europeans have come across information they believe was false or misleading several times a month or more. 85% of respondents perceive this as a problem in their country.
- 83% of Europeans consider fake news to present a problem for democracy in general, either "definitely" (45%) or "to some extent" (38%).
- Respondents to the public consultation considered that disinformation spreads more easily via online media because it appeals to readers' emotions (88%), can influence the public debate (84%), and is designed to generate revenues (65%).

The spread of both disinformation and misinformation can have a range of harmful consequences, such as threatening our democracies, polarising debates, and putting the health, security and environment of EU citizens at risk. Large-scale disinformation campaigns are a major challenge for Europe and require a coordinated response from EU countries, EU institutions, online platforms, news media and EU citizens. The Commission has developed a number of initiatives to tackle disinformation:

- the Communication on 'tackling online disinformation: a European approach' is a collection of tools to tackle the spread of disinformation and ensure the protection of EU values;
- the Action plan on disinformation aims to strengthen EU capability and cooperation in the fight against disinformation;
- the European Democracy Action Plan develops guidelines for obligations and accountability of online platforms in the fight against disinformation;
- The 2018 Code of Practice on disinformation was the first time worldwide that industry has agreed, on a voluntary basis, to self-regulatory standards to fight disinformation. It aimed at achieving the objectives set out by the Commission's Communication presented in April 2018;
- the COVID-19 disinformation monitoring programme, carried out by signatories of the Code of Practice, acted as a transparency measure to ensure online platforms' accountability in tackling disinformation;
- EDMO is an independent observatory bringing together fact-checkers and academic researchers with expertise in the field of online disinformation, social media platforms, journalist driven media and media literacy practitioners;
- the Strengthened Code of Practice on Disinformation, signed on 16th June 2022, brings together a wide range of players to commit to a broad set of voluntary commitments to counter disinformation.

2.1 Initiatives/Communication

The Communication outlines the key overarching principles and objectives which should guide actions to raise public awareness about disinformation and tackle the phenomenon effectively, as well as the specific measures which the Commission intends to take in this regard. In the Commission's view, the following overarching principles and objectives should guide action to tackle disinformation:

- First, to improve transparency regarding the origin of information and the way it is produced, sponsored, disseminated and targeted in order to enable citizens to assess the content they access online and to reveal possible attempts to manipulate opinion.
- Second, to promote diversity of information, in order to enable citizens to make informed decisions based on critical thinking, through support to high quality journalism, media literacy, and the rebalancing of the relation between information creators and distributors.
- Third, to foster credibility of information by providing an indication of its trustworthiness, notably with the help of trusted flaggers, and by improving traceability of information and authentication of influential information providers.
- Fourth, to fashion inclusive solutions. Effective long-term solutions require awareness-raising, more media literacy, broad stakeholder involvement and the cooperation of public authorities, online platforms, advertisers, trusted flaggers, journalists and media groups.

2.2 Initiatives/Action Plan

The Action Plan against disinformation: a set of actions aiming to build up capabilities and strengthen cooperation between Member States and EU institutions to proactively address disinformation.

2.2.1 Improve Detection, Analysis and Exposure of Disinformation

- Invest in digital tools, data analysis skills and specialised staff within the EU institutions and in Member States.
- Assess reach and impact of disinformation.

Stronger cooperation and joint responses to disinformation

- Rapid Alert System consisting of EU institutions and EU Member States set up
- Rapid Alert System to provide alerts on disinformation campaigns, share insights and facilitate coordinated response.

Mobilize private sector to tackle disinformation

- Implementation of EU-wide Code of practice on disinformation
- Major internet companies such as Google, Facebook, Twitter and Mozilla have signed up
- Regular reporting since January 2019 and possible regulatory action in case of unsatisfactory results.

Raise awareness and improve societal resilience

- Targeted campaigns in Europe and beyond
- Active participation of civil society in identifying and exposing disinformation
- Supporting independent media and fact-checkers.

3 Libraries

Considering libraries as places of valid knowledge, it requires librarians to strengthen critical thinking in recognizing not only the value of the content of information but also evaluating the most effective model of Informational Education based on its design process.

3.1 De Paor and Heravi (2020)

De Paor and Heravi (2020), studied how the field of librarianship can help combat the phenomenon of fake news and disinformation. They focused on analyzing current and emerging information literacy contexts and the efforts and initiatives libraries are taking to counter the spread of fake news and educate communities on how to navigate in an age of truth and misinformation. Their findings were focused

- in identifying fake news
- in the analysis of contemporary and emerging contexts of information literacy
- to the efforts and initiatives undertaken by libraries to counter the spread of fake news and educate communities about how to navigate an age of truth and misinformation.

3.2 Lim (2020)

A pilot study, examined how librarians understand the term fake news and their approaches to detecting fake news or evaluating news sources through guides in academic libraries. This paper also provides (Lim, 2020):

- a new definition of fake news, which helps us consider multiple layers of the phenomenon of contemporary fake news (misinformation and disinformation, are frequently used or preferred when referring to fake news)
- a few suggestions:
  - Librarians must further clarify the term fake news so that it reflects its multiple layers.
  - Librarians must incorporate new strategies, such as lateral reading and click restraints, in combination with a few prioritized elements of a checklist into their guides regarding detecting fake news.
  - Librarians must pay attention to psychological factors more when interpreting facts in their strategies about news sources and fake news.

References

Berkowitz, M.B.: ResearchGate (1990). Retrieved from: Information Problem-Solving: The Big Six Skills Approach: https://www.researchgate.net/publication/234713449_Information_Problem-Solving_The_Big_Six_Skills_Approach

Bruce, H.: ResearchGate (2004, January). Retrieved from: Information Behavior that Keeps Found Things Found: https://www.researchgate.net/publication/26395923_Information_Behavior_that_Keeps_Found_Things_Found

Commission, E.: EC Shaping Europe's digital future (2022, 29 June). Retrieved from: EC: https://digital-strategy.ec.europa.eu/en/policies/online-disinformation

Gallagher, B.: What impact does fake news have on the real world? Mary Jane (2016). Retrieved from: https://merryjane.com/culture/fake-news-internet-impact

De Paor, S., Heravi, B.: Science Direct. Retrieved from: Information literacy and fake news: how the field of librarianship can help combat the epidemic of fake news (2020, 5 Sept). https://www.sciencedirect.com/science/article/pii/S0099133320301099?via%3Dihub

Korobili, S., Togia, A.: Theories mathisis (2016). Retrieved from: http://hdl.handle.net/11419/2704

Lim, S.: Science Direct. Retrieved from: The Journal of Academic Librarianship (2020, 5 Sept). https://www.sciencedirect.com/science/article/abs/pii/S0099133320300987

Brand-Gruwel, S., Wopereis, I., Walraven, A.: University of Twente. Retrieved from: Research Information—a descriptive model of information problem solving while using internet (2009). https://research.utwente.nl/en/publications/a-descriptive-model-of-information-problem-solving-while-using-in

Anunobi, C., Udem, O.K.: ResearchGate. Retrieved from: Information literacy competencies: a conceptual analysis (2014). https://www.researchgate.net/publication/283852376_Information_Literacy_Competencies_A_Conceptual_Analysis

Services to Users' Groups

Responding to Labour Market Needs in Industrial Property Education in Slovenia and Croatia. How Can Academic Libraries Contribute?

Tjaša Jug[1](✉), Polona Vilar[1], Katarina Švab[1], Denis Kos[2], Sonja Špiranec[2], and Mihaela Banek Zorica[2]

[1] Faculty of Arts, University of Ljubljana, Ljubljana, Slovenia
{tjasa.jug,polona.vilar,katarina.svab}@ff.uni-lj.si
[2] Faculty of Humanities and Social Sciences, University of Zagreb, Zagreb, Croatia
{dkos,sspiran,mbanek}@ffzg.hr

Abstract. This paper reports on preliminary findings of an ongoing international ERASMUS + research project "IPEDU—Introducing Intellectual Property Education for Lifelong Learning and the Knowledge Economy". The project aims to bolster industrial property (IP) education for librarians, information professionals and staff at technical universities in Europe. To strengthen the role of libraries in IP education, and explore how they can connect with the needs of the economy, present research explores the awareness about IP skills and competences of non-lawyers (engineers, scientists, business managers, etc.) within labor markets and compares preliminary findings of two project partners, Croatia and Slovenia, who have comparable educational contexts. The paper presents results of original research and offers practical and methodological recommendations.

Keywords: Industrial property · Education · Croatia · Slovenia · Labor market needs · Academic libraries

1 Introduction

This paper reports on preliminary findings of an ongoing international ERASMUS + research project "IPEDU—Introducing Intellectual Property Education for Lifelong Learning and the Knowledge Economy". The IPEDU project is led by the Technological University of the Shannon: Midlands Midwest (Ireland) in cooperation with the project consortium (see [3]). The project aims to make expert interventions in current educational programs for librarians, information professionals and staff at technical universities in Europe to bolster industrial property (IP) education. There is a demand in the labor market for workers who have a high level of awareness and competencies in the IP field [8]. Academic librarians are slowly transforming from mere instructors to active collaborators in the educational process [1]. Many academic libraries today employ a copyright expert (Schmidt & English 2015), or a similar special professional

A. Katsirikou (Ed.): QQML 2023, 68, pp. 67–84, 2026.
https://doi.org/10.1007/978-3-032-01264-7_6

profile. Interestingly, most of these profiles have a degree or background in Library and information science rather than formal legal education [6]. However, currently there seems to be little evidence that librarians are recognized as partners and contributors to the development of competencies and skills in the field of IP rights. The research presented here aims to address this research gap by exploring the awareness about IP skills and competencies of non-lawyers (engineers, scientists, business managers, etc.) and their position within labor markets. Therefore, an investigation into the real needs and expectations of employers in the labor market is necessary. This paper compares the preliminary project findings of two partners, Croatia and Slovenia, who have comparable educational contexts to ascertain the level of awareness and attention of employers to industrial property skills and competencies of non-lawyers. This comparison is natural due to the common historical background of both countries, and similar organization of educational and economic systems. In conclusion the paper offers some practical and methodological recommendations, based on the results of original research, about how to improve the work of higher education institutions, strengthen the role of libraries in IP education and the ways in which they can connect with the needs of the economy in this sector.

2 Methodology

An online survey was compiled by IPEDU team members and the same survey was distributed in all participating IPEDU countries in order to get comparable results. The survey contained 11 menu-based, Likert-scale, and alternative questions asking about various aspects of IP in the workplace. The target population were companies and offices that deal with IP, work with employment candidates, or have an IP department or function, and also employ technical profiles. Although the response rate was low in both countries (19 in Croatia and 50 in Slovenia), responses reflect attitudes and needs of employers regarding IP competencies and indicate possible directions for the involvement of academic librarians in IP education.

Our research questions were:

- RQ1: How important are industrial property skills and competencies of non-lawyers (engineers, scientists, business managers, entrepreneurs) to Croatian and Slovenian employers?
  - Which differences in importance, if any, can be observed in different industry sectors comparatively in Slovenia and Croatia?
  - Which differences in importance, if any, can be observed regarding different industrial property skills and competencies of non-lawyers comparatively in Slovenia and Croatia?
- RQ2: Which industrial property skills and competencies of non-lawyers (engineers, scientists, business managers, entrepreneurs) are respectively needed in the labour market of Slovenia and Croatia?
  - Are various industrial property skills and competencies of non-lawyers perceived differently in terms of importance in Slovenia than in Croatia?

 - Which differences in the perception of industrial property skills and competencies of non-lawyers, if any, can be observed in different industry sectors comparatively in Slovenia and Croatia?

- RQ3: What is the attitude of Slovenian and Croatian employers towards training in industrial property?

(a) Slovenia

IPEDU project partners from Slovenia distributed the survey in several ways in order to attract the attention of potential respondents: through the national Office for intellectual property, national Chamber of commerce and industry, Intellectual property office of the University of Ljubljana, Central Technical Library and various individual channels. Data gathering lasted from February 18, 2021, to October 23, 2021. The survey was clicked on by 215 respondents (of which 94 only clicked on the first page) and resulted in 50 adequately filled-in surveys. The questionnaire was translated into Slovenian and prefaced with a short introduction of the project, the objectives of the research and a privacy statement.

(b) Croatia

Of intellectual property-intensive industries in Croatia (industries that make intensive use of the different types of intellectual and industrial property rights). As such, purposive sampling, i.e. a form of nonprobability sampling that involves the selection of specific parts of the population was used to collect data relevant to this research. The sampling frame for this study was an openly available platform created by the Croatian Chambers of Economy, an umbrella institution of the Croatian economy. The platform, Digital Chamber, provides insight into the operations of all business entities on the territory of the Republic of Croatia, including general information, financial indicators etc., and provides access to a search engine indexed by industry branches, based on a national classification of activities (NKD—National Classification of Activities).

Since participation in the survey depended on the voluntary decisions of respondents at companies included in the sample, there was no selection bias involved in who participated in the study. Data collection was conducted for two weeks in February 2021. Using the platform 307 companies were approached, however only 19 filled out the online questionnaire. The questionnaire was translated into Croatian and prefaced with a short introduction of the project, the objectives of the research and a privacy statement (Fig. 1).

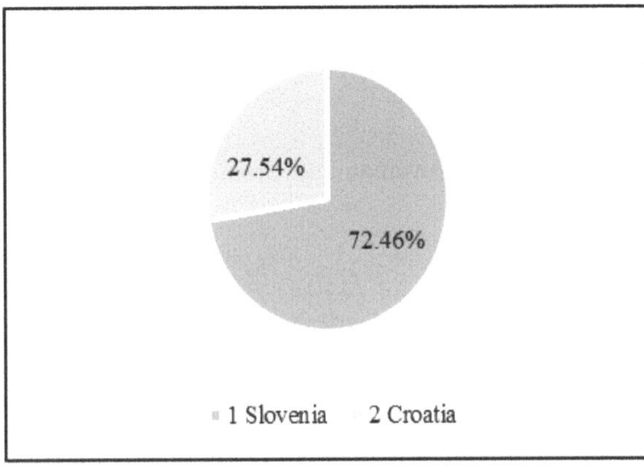

Fig. 1. Participants country of origin

3 Results

The results of the survey are presented here using descriptive statistics by comparing the relative values of Slovenian and Croatian samples. Each figure is interpreted in terms of stated research questions with a focus on comparative analysis between the two analyzed countries.

We start our analysis by comparing the samples in terms of the industry sectors our respondents come from. From Fig. 2. it is clear that the Croatian sample is biased towards the engineering and technology sector while the larger, Slovenian sample shows a more varied structure. This discrepancy can be taken as an indication that the interest for industrial property protection in this sector in Croatia is higher than in other parts of industry.

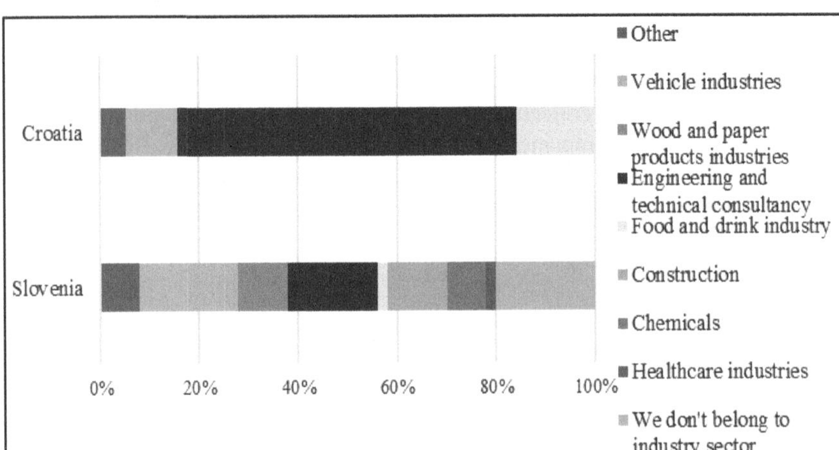

Fig. 2. Sample differences in covered industry sectors

In order to truly explore this interest, we can see another difference (Fig. 3) in Croatian and Slovenian samples. It seems that Croatian respondents consider industrial property knowledge very relevant to their endeavors, while the Slovenian sample shows a more moderate interest with a third of respondents claiming that industrial property rights knowledge is not relevant to them at all. Interestingly, there were no respondents in Croatia who dismissed its importance altogether.

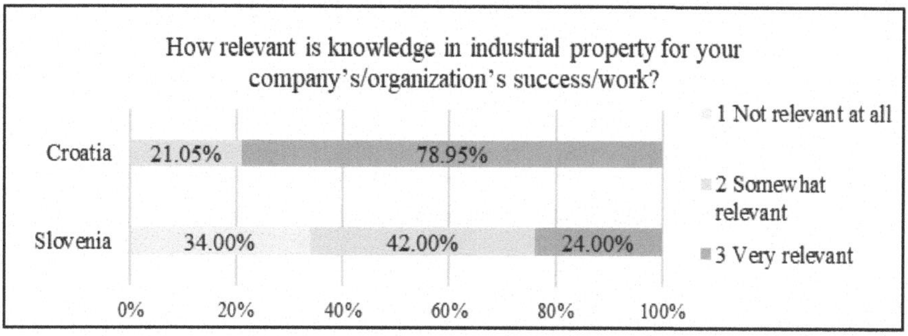

Fig. 3. Relevance of IP knowledge for studied companies/organizations

Some answers may be provided if we consider the frequency in which these companies actually deal with industrial property issues (Fig. 4). Again, two thirds of the Croatian sample express either a moderate or high frequency which is congruent with their heightened interest for these topics. Similarly, almost 60% of the Slovenian sample state that they very rarely deal with industrial property issues which might explain why these respondents find industrial property rights knowledge less relevant for their work.

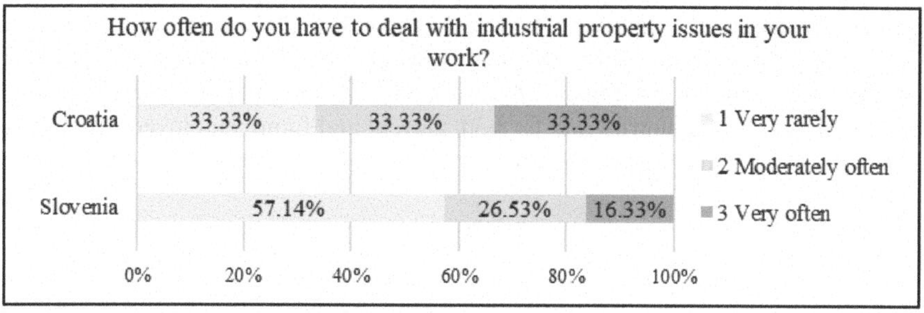

Fig. 4. Frequency of IP issues

A clear statement of interest for industrial property knowledge was gained by asking employers about the potential benefit their company would gain if their employees had industrial property knowledge. Here again we see that Croatian respondents see industrial property knowledge as more beneficial than Slovenian respondents do. This difference might be due to the industry sectors covered in both studies. Thus, in relation to RQ1, we conclude that industrial property knowledge is at least somewhat relevant to employers in both countries, but the discrepancies in covered industry sectors also indicate that these opinions may vary extremely in different parts of industry (Fig. 5).

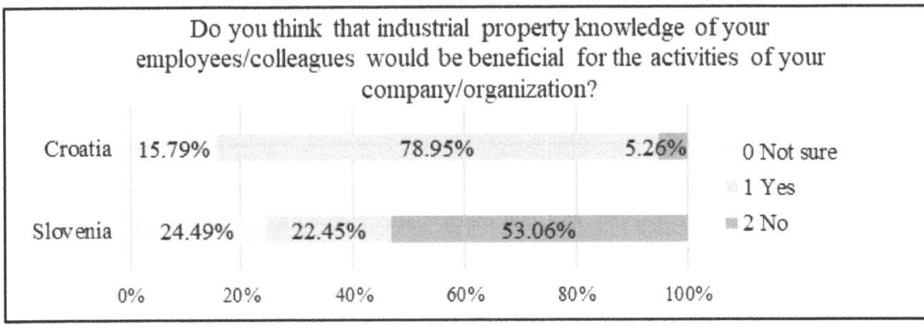

Fig. 5. Employers perceived benefit of IP knowledge

In Figs. 6, 7, 8, 9, 10, 11, 12, 13, 14, 15, 16, 17, 18 and 19 we focus on exploring the RQ2 and the value attributed by employers to specific industrial property skills and competencies. Each skill and competency are considered in a two-fold way i.e. as related to employers' perception of their importance for future and current employees. To start we asked employers whether knowing what industrial property is (Fig. 6) has relevance. We can see that Slovenian responses remain similar both in the case of job candidates and current employees. However, in case of Croatia employers perceived that knowing what is industrial property is more important for current employees then it will be for future employees. Such differences are not observed when it comes to knowledge about industrial property law; how to utilize industrial property knowledge in their work; where to find information on existing patents, trademarks, models etc.; knowledge regarding the process of obtaining protection of industrial property; knowledge of industrial-property related risk management; ability for financial assessment of industrial property assets; nor for knowledge how to utilize the industrial property domain in order to acquire information on competitors (Figs. 7, 8, 9, 10, 11, 12, 13 and 14). Here the answers remain stable for both countries and for both current and future employees, but we can establish that Croatian respondents attribute a somewhat higher importance across all skills and competencies.

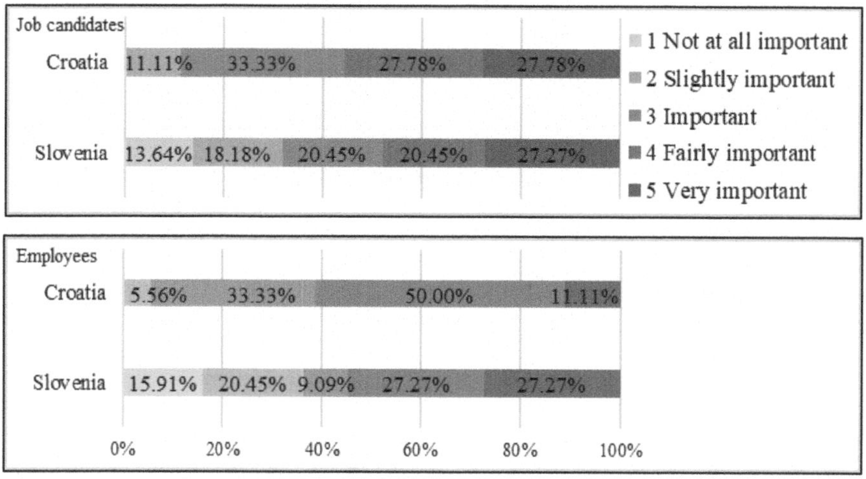

Fig. 6. Knowledge what IP is

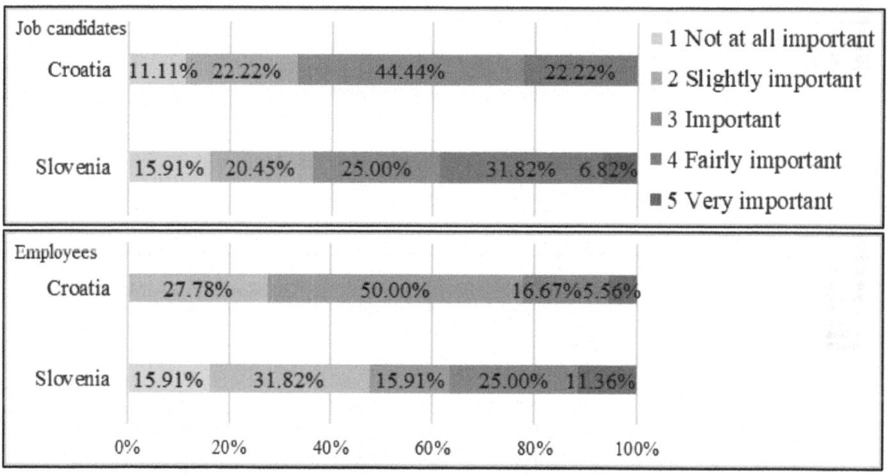

Fig. 7. Knowledge of the basics of industrial property law

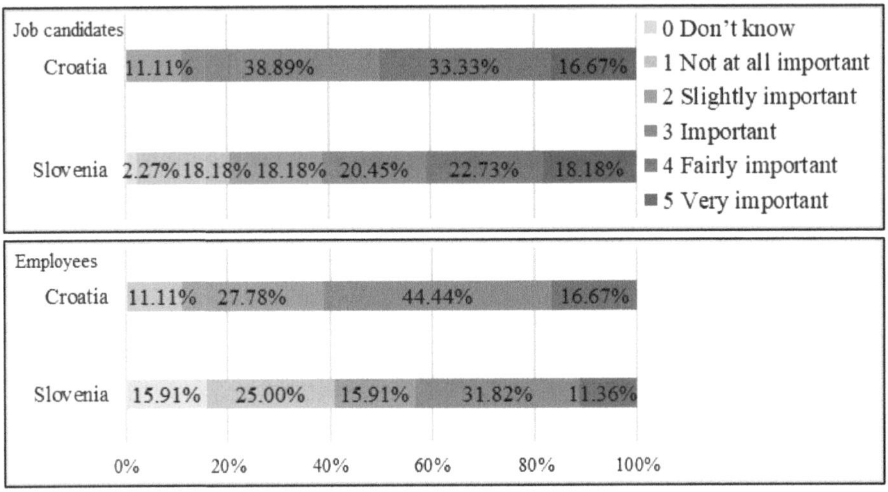

Fig. 8. Knowledge how to utilize industrial property in their work

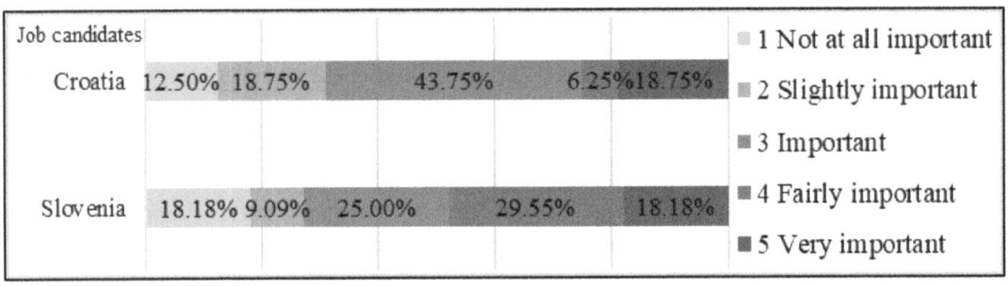

Fig. 9. Knowledge how to find information on patents, trademarks, models—job candidates

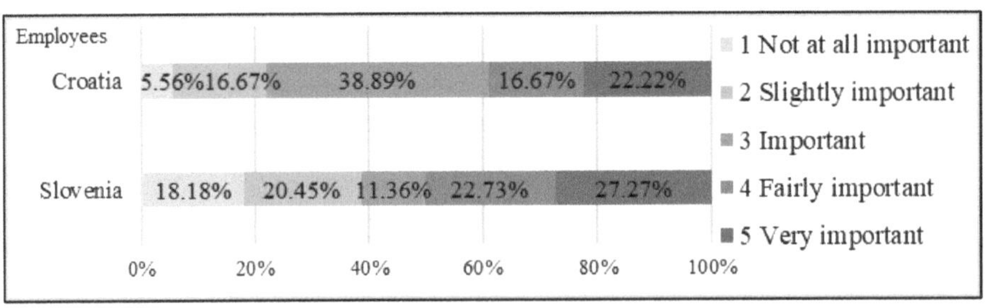

Fig. 10. Knowledge how to find information on patents, trademarks, models—employees

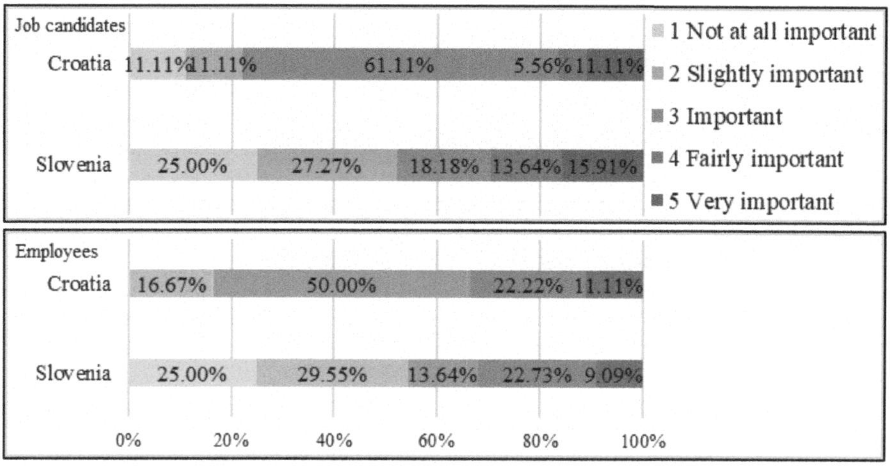

Fig. 11. Knowledge regarding obtaining protection of industrial property

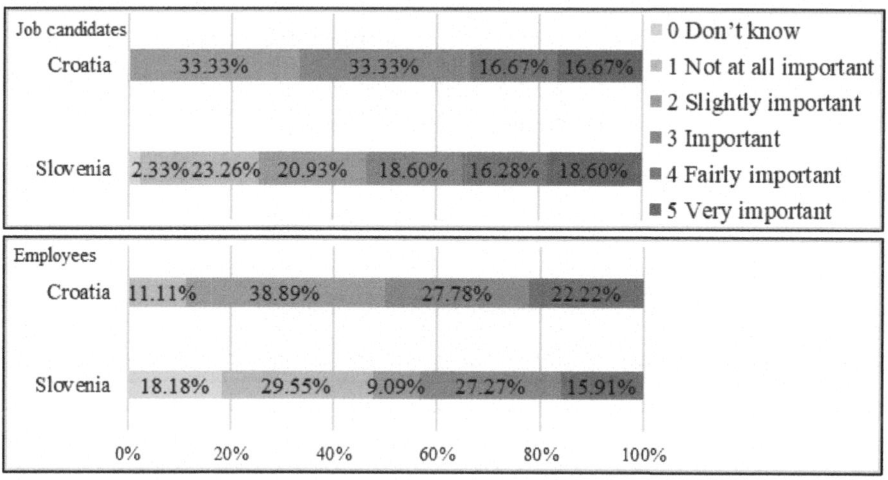

Fig. 12. Knowledge of industrial-property-related risk management

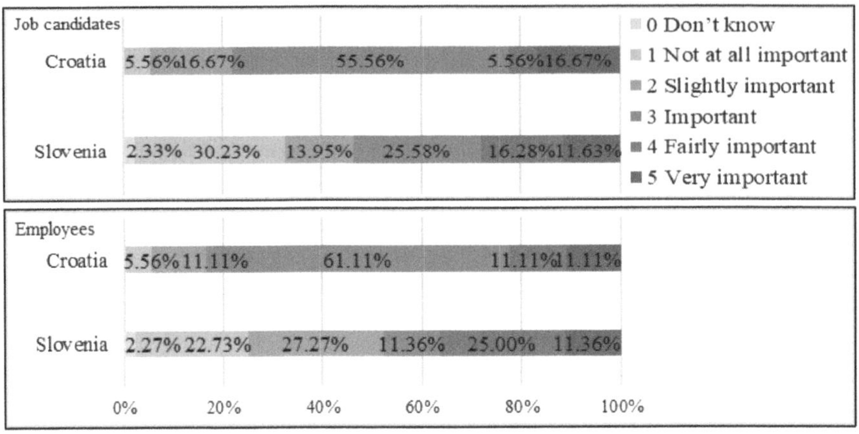

Fig. 13. Ability for financial assessment of industrial property assets

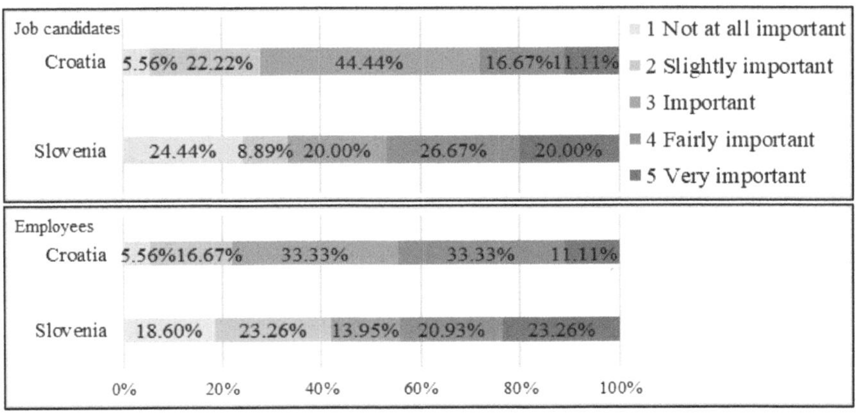

Fig. 14. Knowledge how to utilize the industrial property domain in order to acquire information on competitors

Among the skills and competencies which were attributed high importance, agreement between the studied countries exists in relation to knowledge of hazards of infringing or misappropriating other people's industrial property assets (Figs. 15 and 16). More ambiguous answers leaning towards medial values occur for the rest of the studied skills and competencies as can be seen in Figs. 17, 18, and 19.

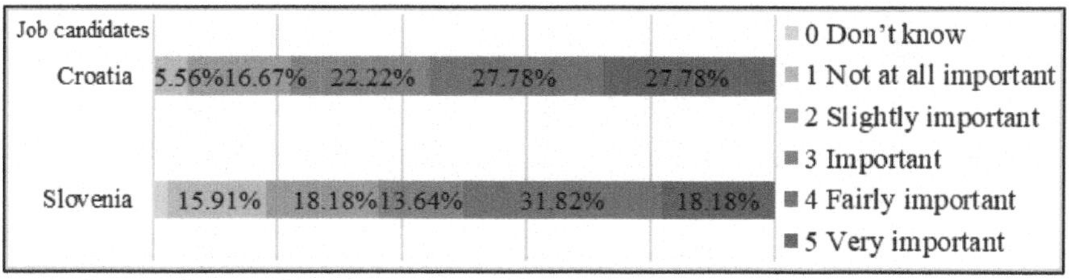

Fig. 15. Knowledge of hazards of infringing or misappropriating other people's industrial property assets—job candidates

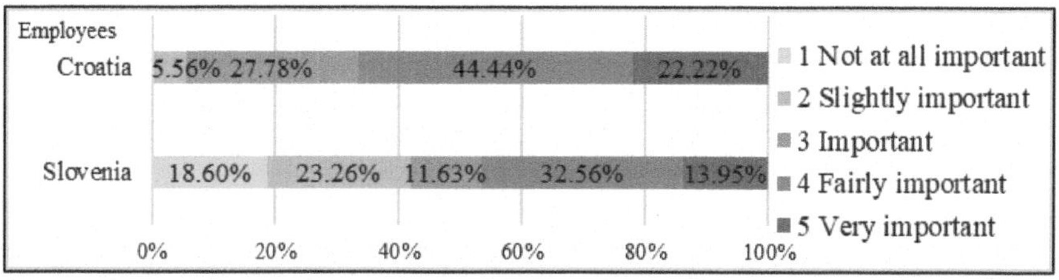

Fig. 16. Knowledge of hazards of infringing or misappropriating other people's industrial property assets—employees

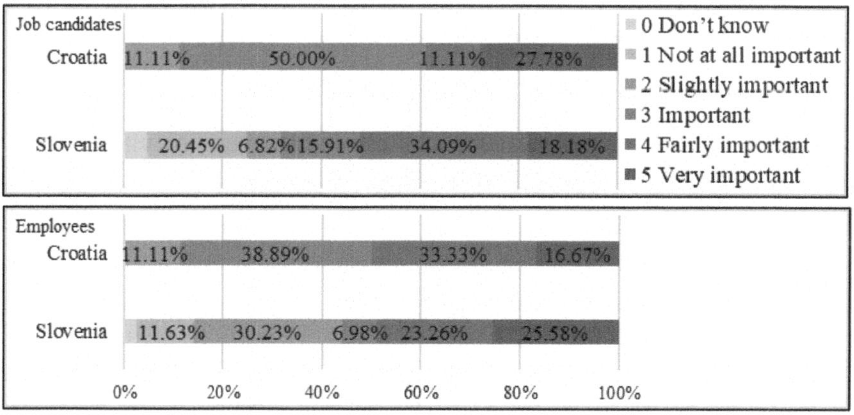

Fig. 17. Awareness when and how to consult about industrial property

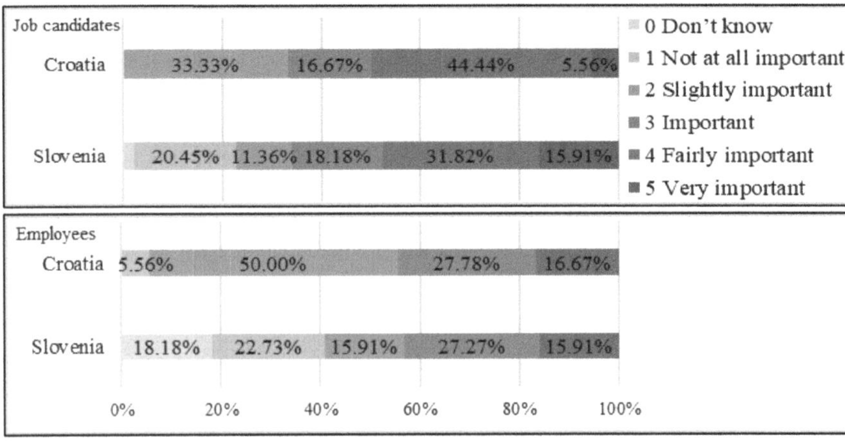

Fig. 18. Basic knowledge on the different types of intellectual property

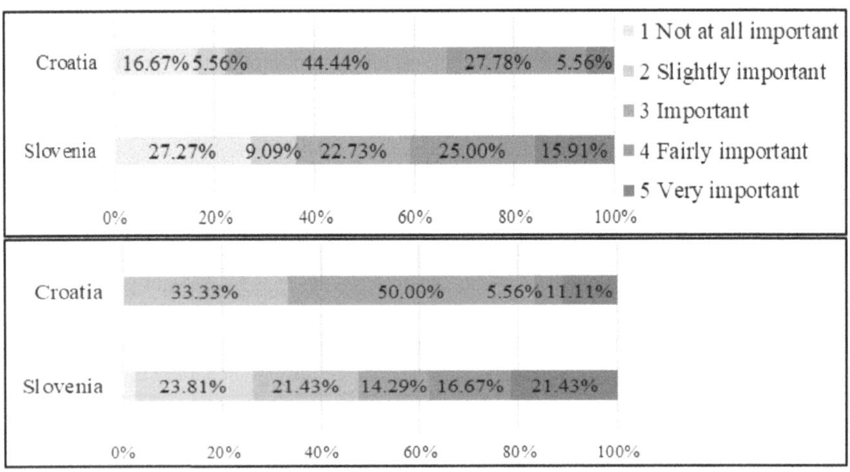

Fig. 19. Knowledge of the role industrial property assets play in strategic business planning

On average and in comparison, the most and least valued skills and competencies for Croatian and Slovenian employers are represented in Table 1.

Table 1. Comparison of importance attributed to IP skills and competencies by employers

		Croatia	Slovenia
Job candidates	Most important	• Knowledge what IP is • Knowledge of hazards of infringing or misappropriating other people's industrial property assets • Knowledge on how to utilize industrial property in their work • Knowledge of the role industrial property assets play in strategic business planning	• Knowledge what IP is • Knowledge of hazards of infringing or misappropriating other people's industrial property assets • Knowledge where to find information on existing patents, trademarks, models etc • Awareness when and how to consult with an industrial property expert or adviser
	Least important	• Knowledge of industrial-property-related risk management	• Ability for financial assessment of industrial property assets
Employees	Most important	• Knowledge what IP is • Knowledge of hazards of infringing or misappropriating other people's industrial property assets • Knowledge how to utilize industrial property in their work	• Knowledge what IP is • Knowledge of hazards of infringing or misappropriating other people's industrial property assets • Knowledge how to utilize industrial property in their work
	Least important	• Knowledge of the basics of industrial property law	• Knowledge of the basics of industrial property law

In response to RQ2, Table 1 shows that a clear agreement among Croatian and Slovenian employers exist that knowing what industrial property is and being able to discern the hazards related to rights infringement and misappropriation of other people's assets are key concerns and most valued competencies. These are followed by employers' common interest that their employees, both current and future, should know how to utilize industrial property in their work.

Finally, to consider RQ3, attitudes towards industrial property education and training were explored. The results (Fig. 20) show that around 40% of employers in both Slovenia and Croatia offer training in industrial property at least sometimes. However, a concerning discovery is the reverse i.e. that over 55% of employers from both countries never or rarely offer such trainings.

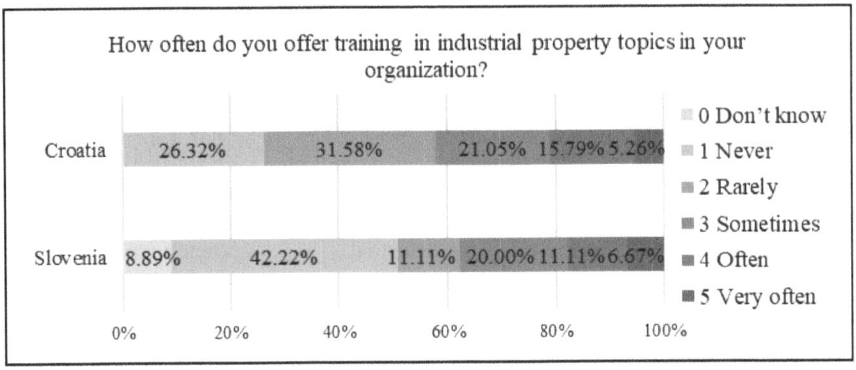

Fig. 20. Frequency of training offers in industrial property for non-lawyer employees

However, a key difference can be observed between the two studied countries when we asked whether such training would be beneficial for work and success of respondents' companies (Fig. 21). Interestingly, almost 58% of Croatian employers think they would be beneficial, while over 50% of Slovenian employers think such training has no merit for the success and work of their companies.

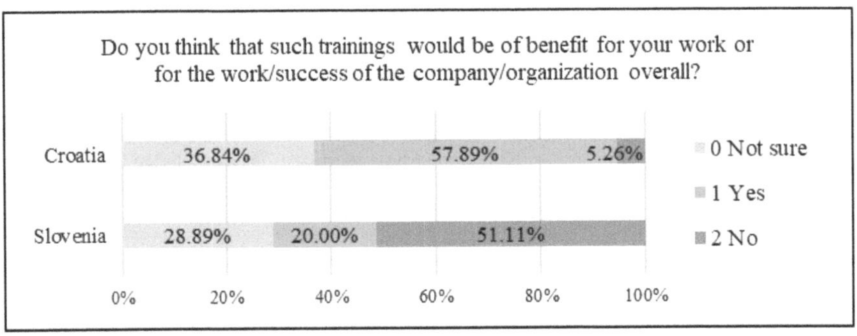

Fig. 21. Employers' perceived benefit of industrial property education and training offers

When asked how often these trainings should be organized (Fig. 22), again, Croatian employers opt for a stable frequency of such educational offerings and are less interested in reactive educational strategies. Slovenian employers are more ambiguous with relatively even distribution of opinions on the scale.

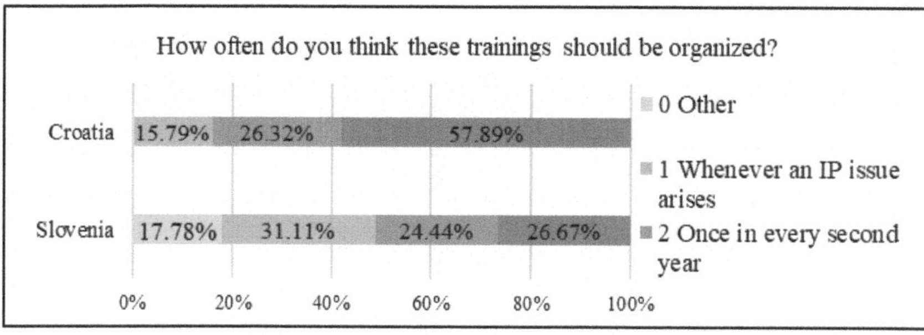

Fig. 22. Employers' opinion about needed frequency of training industrial property topics

We also inquired about what kind of expertise employers would prefer to rely on when dealing with industrial property issues (Fig. 23). While Croatian respondents indicate a preference of having an in-house specialist trained in the field of industrial property, Slovenian respondents prefer hiring outside industrial property specialists. There is too little information to determine why that is so, and certainly future studies could explore the influences which have shaped the differences in these stated preferences.

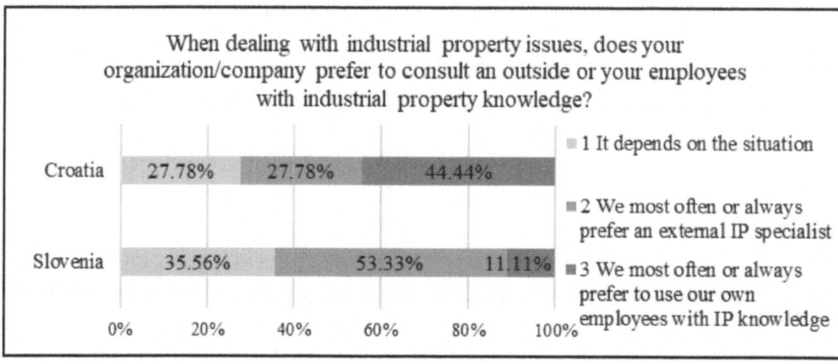

Fig. 23. Employers' preference in choice of experts to resolve industrial property related issues

Finally, opinions about the benefit of having employees with industrial property knowledge were investigated (Fig. 24). Here Slovenian employers again show a more negative attitude towards industrial property knowledge and education, but the results are more congruent between the countries. However, a significant percentage of studied employers state that they are unsure whether having such employees would be advantageous for company interests. The question arises who should then be those who offer industrial property education and training if this knowledge is somewhat valued but not readily offered to company employees.

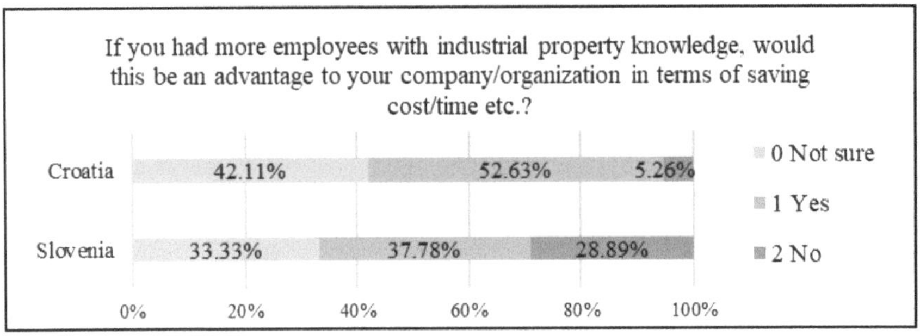

Fig. 24. Employers' perceived value of employees with industrial property knowledge and skills

4 Discussion, Methodological and Practical Recommendations

Acknowledging low response rates, results show that employers value IP competencies of their employees and acknowledge their positive impact, but Slovenian employers don't seem to perceive them as decisive for the success of their company. This is an interesting finding if we consider the World Intellectual Property Organization (i.e. WIPO) position on the necessity of intellectual property education for non-jurists. It seems that the industry sector has yet to agree with the WIPO's assessment that intellectual, and hence, industrial property education of non-jurists is important. WIPO's argumentation states that such knowledge will "give them the tools to know when to apply for IP protection and how to manage the creative process without ruling out such protection" and that "it ought to provide them with the ability to know when they may use others' materials and when they may not." [9, p. 12]. However, when asked what competencies employers value the most WIPO's assessment seems to be on point and congruent with employers' answers. This points us to the need to better promote the purpose and value industrial property education of non-jurist employees has for employers in specific industry sectors. Moreover, further studies are needed that would look into the opinions of Croatian employers from a more diverse set of industry sectors. In addition to that, low response rates indicate that future studies may need to develop specific motivational strategies for employers to participate and express their opinions about industrial property knowledge and education.

The results highlight that on average there are many similarities regarding the types of IP-related competencies Croatian and Slovenian companies value. Being able to define and utilize industrial property as well as taking care of avoiding any infringements is what is deemed essential. However, both countries have stated that being knowledgeable about industrial property law is the least important competency. This information could be valuable when designing industrial property training events in terms of program content and the level of expert legal language educators will include in training design and delivery. WIPO [9, p. 12] warns though, that "principles of intellectual property should remain intact", regardless of who the audience being taught is. Furthermore, Kaplan and Kaplan [4] assert that engineers and scientists especially should have basic knowledge of industrial property law because this will allow them to internalize industrial property principles in their everyday work. Our data shows that more advocacy of the value of such knowledge is needed across all industry sectors but also in the higher education arena. This kind of knowledge has been recognized by the European Commission [2] as key in economic exploitation of intellectual property assets that result from creative work in all walks of business and entrepreneurships.

In relation to industrial property educational offers a common conclusion confirming earlier studies [cf. 5, 7] is that employers value organized training events but these are organized infrequently and inadequately. In Slovenia outside specialists are preferred when dealing with industrial property issues, while in Croatia more emphasis is on having in-house experts. However, the answers pointing to the uncertainty in estimation of the benefit employees knowledgeable in industrial property bring to companies show that employers might not see themselves as primary disseminators of that knowledge. This role, for now, remains the role of higher education institutions. In them, we stipulate, libraries should be recognized as valuable partners and support in dealing with industrial property issues. Special and academic libraries as documentation centers often function as hubs of pertinent and specialized information resources students learn to navigate. They are initiated into specifics of the knowledge domains they study and knowing what kind of professional outputs they will be able and are expected to create is crucial (e.g. patents). These creative and intellectual outputs should be presented to them as protectable and marketable assets. Libraries as gate keepers of professional knowledge are naturally positioned to be those that transfer basic knowledge about intellectual property rights protection of students' original works and the works of others, they (will) use in diverse professional contexts. As librarians themselves thus become trainers, higher education institutions should ensure that at least some librarians at technical universities have basic knowledge of industrial property law and that they are seen as embedded partners supporting delivery of industrial property education and training. Libraries should also capitalize on the basis of their traditional roles and functions. By reaching out to diverse industries and treating them as members of their service communities' libraries could answer particular information needs in creative processes of companies. This presents unique opportunities to pass on the beneficial logic of industrial property protection. As public spaces libraries should work on becoming places of business incubation by providing space, resources, technology and support for entrepreneurs and start-ups. Industrial property education fits well with the mission of libraries to support the professional development of citizens in all types of entrepreneurial endeavors. Thus, libraries become an emergent partner in local economic development.

5 Conclusions

The present study explored employers' attitudes towards industrial property knowledge and education in two national contexts—Croatia and Slovenia. As these countries share a common historical background their educational and economic contexts are comparable. Through a comparative approach we explored the importance of industrial property knowledge for employers, their preferences and attitudes towards specific industrial property skills and competencies, as well as attitudes towards industrial property educational and training offers. Having explained methodological challenges in sparking the interest of employers to participate in the study key sample differences showed some differences in the covered industry sectors in both studied countries respectively.

These differences showed that different industry sectors may perceive industrial property knowledge in qualitatively and substantially different ways. These differences may stem from cultural, economic and business contexts of studied countries as well as diverse company profiles and interests, but more research is needed to uncover the true nature of these differences. The value attributed to specific competencies did not show substantial differences, but least valued competencies which pertained to knowledge of industrial property law showed that employers themselves might need some kind of awareness raising regarding these topics. General value of industrial property education and training was recognized but existing offers need to be further developed and their value promoted in both industry and higher education settings. Finally, the role of libraries was emphasized and discussed, since as gate keepers to professional knowledge, libraries and librarians are usefully positioned to engage in advocacy and industrial property education and training.

References

1. Corrall, S., Jolly, L.: Innovations in learning and teaching in academic libraries: alignment, collaboration, and the social turn. New Rev. Acad. Librarianship **25**(2–4), 113–128 (2019)
2. European Commission. Directorate General for Research.: *Investing in Research: An Action Plan for Europe: Communication from the Commission to the Council, the European Parliament, the European Economic and Social Committee and the Committee of the Regions* (vol. 226). Office for Official Publications of the European Communities (2003). Available at: http://eurlex.europa.eu/smartapi/cgi/sga_doc?smartapi!celexplus!prod!DocNumber&lg=en&type_doc=COMfinal&an_doc=2003&nu_doc=226
3. IPEDU.: IPEDU. Last modified 24.09.2022 (2021). Available at: https://www.ipeduproject.eu/
4. Kaplan, K.M., Kaplan, J.J.: Incorporating intellectual property into engineering education. In: *Proceedings of the 2003 American Society for Engineering Education Annual Conference and Exposition* (2003). Available at: http://ipmall.info/hosted_resources/Teaching_IP/Kathleen_Kaplan_2003.pdf
5. Pattanapairoj, S., Nitisiri, K., Sethanan, K.: A gap study between employers' expectations in Thailand and current competence of master's degree students in industrial engineering under industry 4.0. Prod. Eng. Arch. **27**(1), 50–57 (2021). https://doi.org/10.30657/pea.2021.27.7
6. Patterson, E.: The Canadian university copyright specialist: a cross-Canada selfie. Partnership: Can. J. Libr. Inf. Pract. Res. **11**(2) (2016)
7. Philbin, S.: Entrepreneurship education for engineers: technology evaluation and commercialization through project-based learning, in *American Society for Engineering Management 2021 International Annual Conference and 42nd Annual Meeting*. ASEM (2021)
8. Soetendorp, R.: Teaching intellectual property to non-law students, in Takagi, Y., Allman, L., Sinjela, M. (eds.) *Teaching of Intellectual Property: Principles and Methods*. Cambridge University Press (2008)
9. World Intellectual Property Organization (WIPO).: *Teaching Intellectual Property (IP) in Countries in Transition* (2013). https://doi.org/10.34667/tind.28958

Enhancing the Human Experience: Infopragmatics-Driven Integration of Artificial Intelligence in Digital Libraries

Rafael Ibarra(✉), A. Rendón, D. Ortiz, E. Ávila, Y. Macedo, and S. Algarín

DGBSDI, Universidad Nacional Autónoma de México, Mexico City, Mexico
ibarraraf@aol.com

Abstract. This case study highlights the successful use Infopragmatics as a driven force in integrating AI into UNAM's Digital Library workshops. It details the Infopragmatics steps involved in searching catalogues, specialised academic information databases and discovery services, culminating in using Chat-GPT as a Table of Contents (TOC) advisor and for English improvement in abstracts. The results of this integration are discussed, demonstrating its effectiveness and pertinence in the academic fields of Chemistry, Accountancy and Administration.

Keywords: Infopragmatics · AI · Large language models · LLM · Chat-GPT · Digital libraries · UNAM · Chemistry · Accountancy · Administration

1 Introduction

The objective of this manuscript is to demonstrate the establishment of foundational principles for the integration of Artificial Intelligence into the UNAM's Digital Library using the infopragmatics method. Ibarra's [3] Infopragmatics leverages three applied linguistics techniques: the research sentence (Tlacuilo); the Linguistic storm, and a Controlled Vocabulary Matrix. Through our findings, we identify a range of advantageous outcomes for harnessing Artificial Intelligence, specifically ChatGPT-4, yet our methodology holds promise for application across various other Large Language Models (LLM) as well.

2 Methodology

This study encompasses two distinct groups in which Infopragmatics was implemented: the first group consisted of undergraduate students specialising in Chemistry Science, while the second group comprised graduate students in Accountancy and Administration. Both groups are affiliated with the same university: UNAM. The workshops took place during the periods of February (10 h), and March (20 h).

Both groups participated in the workshop titled: "Assertive Bibliographic Research", wherein they utilized various collections from the Digital Library, including e-Dissertations; e-Books; Commercial Academic Information Systems; e-Journals, and

A. Katsirikou (Ed.): QQML 2023, 68, pp. 85–93, 2026.
https://doi.org/10.1007/978-3-032-01264-7_7

a Discovery Service. However, their manuscripts objectives differed: while undergraduate students aimed to develop dissertations, graduate students aimed to produce research papers. The workshop placed significant emphasis on the application of Infopragmatics principles to conduct efficient bibliographic research tailored to the participants' academic pursuits. Immediately following this emphasis, the participants were guided to:

- Craft a Research Sentence. Tlacuilo technique.
- Translate the sentence from Spanish to English
- Consult available catalogues, starting with the native language and then transitioning to the lingua franca, English.
- Initiate a linguistic storm composed of nine elements, drawing from onomasiological sources such as thesauri, keywords, MeSH, indices, specialised dictionaries or encyclopedias, and ontologies. It is important to note that not all commercial academic information systems include all these resources. We also recommend looking for printed onomasiological resources.
- Expand the Controlled Vocabulary Matrix ($3 \times 3 \times 3$) in the previously explored catalogues [4].
- Once all these steps are completed, it is time to draft a tentative abstract, into three sentences. Begin with the research sentence; then, argument why the selected instrument was utilised, which constituted the second sentence; and conclude with the findings as the third and final sentence. These three sentences together form the abstract. For the title, only an appropriate adjective is needed to adequately modify either the instrument or the phenomenon under study.
- Ultimately, AI serves as the tool to propose a Table of Contents. One may utilize any of the LLM, such as Chat-GPT or others. The results are remarkable. Appendix 2 presents the contrast between the original TOC and the improved one.

3 Results

A.

For the first group the workshop was conducted face-to-face format. Out of the participants, 11 presented their final works. To our surprise as instructors, a group of Chemistry professors attended three of the students' presentations. One of the professors shared an opinion that encapsulated the advantageous utility of infopragmatics. Appendix 1 contains the complete opinion. Here, we present two of the professor's key points:

> I would like to participate in the very workshop myself to improve not only my management as a thesis advisor but also to promote it among my subject professors, especially those whose main occupation is not teaching but industry, Now, more than ever, I objectively realize that we lack the most basic tools to supervise the elaboration of thesis works.
>
> The existence, and above all, the use of countless tools offered by UNAM for bibliographic research are taught. I regret to say that, after 50 years linked to the School of Chemistry, I knew these tools minimally. This not only improves the quality of the content but also greatly helps in the expedited writing of the work.

It is noteworthy that this professor was taken aback by the first presentation and expressed his intention to inquire about the details of each slide once the student concluded his participation.

B.

The workshop for the group of graduate students (91 registered participants) was conducted via Zoom. Their research sentence guided their objectives, which were as follows:

Research papers	28
Dissertations	26
Research papers or Dissertations	3
Did not report advances	34

At the conclusion of the workshop, 15 students presented their final works before the entire group. Appendix 3 shows the original abstract and the improved one. They were asked to express their feelings before and after the workshop. Here, we present two opinions:

1st.

At the beginning, teacher, honestly, I felt insecure because I had some sketches in mind of what could happen, but very insecure regarding the lack of literature of studies and such, that I had within reach, and whether someone else would be interested in that or maybe it's just me because I've been working on the topic for so long. It's relevant to me and no one else. Once you taught us how to consult journals, theses, databases… I realized that it is indeed of interest to many other people and that there is academic literature, and there are studies already conducted previously that will be very useful for me to address a point that, as far as I've gone, is novel and relevant because this type of internal control analysis, and so on, is used in the administrative part. And I want to take it to the judicial power where the topics are predominantly jurisdictional and nobody gets involved; and yet, they are the ones that are very susceptible where staff can fail in integrity and commit acts of corruption. Yes, it can be prevented and also evaluated but we haven't given it much importance. So, I feel more confident now, encouraged. I know where it comes from, that it's a world of information. That's why I say, [to finish my research] 1 year or a little more. But I also know who might find it interesting, [for a] journal and others that I saw. So, I feel happy, satisfied to have taken this course and to have been conducting, as you led us, hand in hand, class by class, each activity, and have my compendium where I will return later when I fully immerse myself in my article.

2nd.

My ideas are as if they were embodied in what I have been working on within the thesis; and, my gratitude. Well, at the level of coordination of the PhD in Administration Sciences that is pushing all these initiatives to, well, take us to how the rest of the universities are on the global level … um… that publish, that many people dedicated to academia… won't hire you if you don't have [publications] in certain journals. So, this is a reflection of a product of academia. The publications, the theses, all those contributions, what you collaborate by supporting others, and what others enrich from your work, right? It's like a genuine interest in the advancement of knowledge. And, well, I believe that having handled this, and as mandatory, is very relevant as was well pointed out and I liked hearing that part from Ángel [another participant] well, whether you are starting, or whether you are in the last semesters of any academic level. I think the workshop was really very good. Thank you very much, thank God for what is behind you… very focused, very, very clear… it takes you… and also it's like you force yourself to work during the class sessions. And, like, um, to have everything ready, right? And, well, to the General Directorate of Libraries of our university. So, basically, that's it, teacher. Thank you very much, and thanks to everyone.

The integrations of infopragmatics-driven AI, specifically Chat GPT, into the Digital Library workshops yielded promising results. Participants reported enhances efficiency in navigating and synthesizing vast amounts of information, leading to more focused and productive research endeavors. Additionally, Chat-GPT's role as a TOC adviser proved invaluable in streamlining the organization of research materials and facilitating the development of coherent academic manuscripts.

4 Discussion

The successful application of Infopragmatics in conjunction with AI underscores the transformative potential of technological advancements in academic research settings. By incorporating AI tools like Large Language Models into the digital library's workflow, universities can empower students and researchers to navigate the increasingly complex landscape of scholarly information effectively. Furthermore, the scalability of Infopragmatics principles suggests broader implications for improving information literacy and research methodologies across diverse academic disciplines.

On the other hand, why should we not use an LLM at the beginning of the workshop? In the first place, libraries provide a guarantee of information reliability, unlike LLMs which may lack such assurance. Besides, LLMs' results, though rhetorical, are out of focus. Moreover, the LLM academic sources are unknown, which represents an unacceptable practice for academic purposes. Additionally, we must pay attention to the Achilles' heel: the hallucinations.

For example, it is worth mentioning what a speaker, Frické [2], asserted during an international conference, convinced of being correct. In his lecture titled "Can librarians be astronauts?", he began by promising significant scientific breakthroughs thanks to artificial intelligence and provided captivating examples.

He continued his lecture: "At times, it can provide information, in text, images, or audiovisual, that is completely biased, false, or mistaken, in addition to lacking transparency." He added that what the Large Language Model (LLM) can do is astounding, and on par with a good, intelligent, and well-informed native speaker of natural language; and that occasionally, it relies on thesauri.

Before bidding farewell to the audience, he assured that the phrase for saying "Thank you" in the Nahuatl language was "tlacuache", according to his LLM. Next, he also demonstrated that the AI assistant, named LLAMA 2 responded to his question about how the phrase was pronounced. The explanation was linguistically accurate, based on the International Phonetic Alphabet.

However, it was pointed out to him by someone in the audience that "tlacuache" is a term that defines a marsupial, and that the phrase for saying 'Thank you' is "Tlazocamati". It is not difficult to discern hallucination between these two terms. Both include the same prefix: "tla". If one were to search for the frequency of each term in any browser, the results would explain why LLAMA 2 discarded the correct option, "tlazocamati", which appears far less frequently than "tlacuache": 31,200 vs 2,910,000.

The utility of AI in various scenarios is undeniable beneficial, as long as it remains free from becoming a tool for indoctrination. As educators, both teachers and librarians, we may concur with Roose [6], in asserting that:

"…schools should thoughtfully embrace ChatGPT as a teaching aid—one that could unlock student creativity, offer personalized tutoring, and better prepare students to work alongside A.I. systems as adults. It is a question of choosing to be critical thinkers or manipulated persons. We still can discern between credible sources and deceptive rhetoric.

Despite advances in ICT (Information and Communication Technology), we are obliged to consider what Cid et al. [1] quote:

"It has been stated that freshmen college students are information illiterate (Henkel et al. 2018; Dawes 2017). This might be due to the belief that reading and writing are skills that people immediately acquire and thus it is unnecessary to teach them at the university (Carlino 2003); which also occurs with IL (Dawes 2017, P. 287).

Furthermore, it is relevant to highlight a contrast between the earlier reference and what Kramer's [5] observation regarding the existence of formally written records around the year 3000 BC:

> No Sumerian writer or scribe, as far as we know, ever made a conscious effort to write a cultural or political history of Sumer or any of its component states, let alone of the then known world. To be sure, Sumerian men of letters originated and developed a number of written literary genres—myths and epic tales, hymns and lamentations, proverbs and essays… But the thought of preparing a connected history, prompted either by the love of learning or even by what we would term purposes of propaganda, never seemed to occur to the Sumerian teachers and writers. P. 37.

5 Conclusions

In conclusion, this manuscript highlights the significant strides made in integrating AI into UNAM's Digital Library through the lens of Infopragmatics. By adopting a systematic approach to information retrieval and processing, coupled with the capabilities of AI technologies like the Large Language Model utilised in this paper, institutions can revolutionize the way knowledge is accessed, synthesized, and disseminated within academic communities. Moving forward, further research and collaboration are warranted to explore the full potential of Infopragmatics-driven AI in advancing scholarly inquiry and enhancing the human experience in the digital age. It is crucial to highlight the persistent presence of both generational gaps among individuals and technological disparities in technology. Neglecting to address these gaps appropriately could lead to undesirable outcomes. While the consequences of LLMs' hallucination might not be detrimental in certain academic domains, relying solely on AI-generated manuscripts to manage hazardous chemical substances could pose serious risks, potentially resulting in lethal outcomes. Therefore, exercising caution and implementing appropriate measures are paramount.

Appendix 1

The following is the email addressed to the librarian responsible for the School of Chemistry, at the UNAM.

Good afternoon, Miss "T".

Today I attended the presentation to which we were kindly invited and witnessed three presentations by students of the workshop.

Allow me to send you my comments which are as follows:

1. I would like to participate in the very workshop myself in order to improve not only my management as a thesis advisor but also to foster it among my subject professors, especially those whose main occupation is not teaching but industry. Now, more than ever; I objectively realise that we lack the most basic tools to supervise the elaboration of thesis works.
2. The content of the workshop not only provides tools but also offers the necessary motivation for writing the work. The absence of this motivation is preventing us from titling more interns. Having a guide on "how to do and organise" the content allows interns to verify their progress, which always motivates them to continue and complete the work.
3. The existence, and above all, the use of the countless tools offered by the UNAM for bibliographic research are taught. I regret to say that, after 50 years linked to the School of Chemistry, I knew those tools minimally. This workshop not only improves the quality of the content but also greatly helps in the expedited writing of the work.
4. It also teaches how to write objectives, chapters, etc. This is an obstacle for the intern to start writing, since it is not formally taught in the undergraduate program. The three presentations we witnessed seemed remarkable to me for their order, conciseness, objectivity, and clarity. Therefore, the usefulness of the workshop is confirmed. It would be worth opening it not only to interns but also to students from the fifth or sixth semester, perhaps with some modification so as not to require the prerequisite of having or being able to identify "a thesis topic", but rather thinking about the internships, reports, and daily tasks about which all teachers constantly complain.
5. I see as an opportunity, although it was explained to me that you are already planning a related workshop. The great advantage of knowing how to access a system that provides countless data allows for obtaining many books, articles, and sources of information. However, if users are not concurrently taught how to organise references and correctly include them in their written work after researching, this information

will be lost at best. At worst, it could lead to issues of improper citation or unintentional plagiarism.

In summary, I think it is an excellent workshop.

Thank you very much for the invitation.

Best regards.

Appendix 2

Original TOC

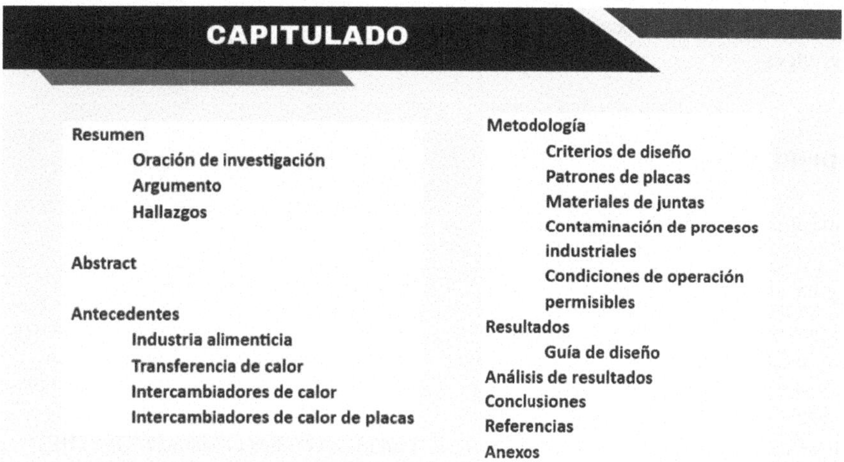
CAPITULADO

Resumen
- Oración de investigación
- Argumento
- Hallazgos

Abstract

Antecedentes
- Industria alimenticia
- Transferencia de calor
- Intercambiadores de calor
- Intercambiadores de calor de placas

Metodología
- Criterios de diseño
- Patrones de placas
- Materiales de juntas
- Contaminación de procesos industriales
- Condiciones de operación permisibles

Resultados
- Guía de diseño

Análisis de resultados

Conclusiones

Referencias

Anexos

TOC Improvement

1. Literature Review
2.1 Overview of Plate Heat Exchangers
2.2 Applications of Plate Heat Exchangers in the Food Industry
2.3 Design Considerations for Plate Heat Exchangers
2.4 Previous Studies on Plate Heat Exchanger Design
2.5 Recent Advances in Plate Heat Exchanger Technology
3. Methodology
3.1 Data Collection Sources
3.2 Inclusion and Exclusion Criteria
3.3 Search Strategy
3.4 Data Extraction and Analysis
4. Design Alternatives for Plate Heat Exchangers
4.1 Traditional Plate Heat Exchanger Designs
4.2 Emerging Trends in Plate Heat Exchanger Design
4.3 Comparative Analysis of Design Alternatives

5. Case Studies
5.1 Case Study 1: Application of Plate Heat Exchangers in [Specific Food Industry Process]
5.2 Case Study 2: Optimization of Plate Heat Exchanger Design for [Specific Food Industry Application]
6. Discussion
6.1 Key Findings
6.2 Implications of Design Alternatives
6.3 Recommendations for Future Research
7. Conclusion
7.1 Summary of Findings
7.2 Contributions to Knowledge
7.3 Practical Implications
7.4 Areas for Future Research
References

Appendix 3

Abstract improvement.

Original abstract	Improved version
The purpose of this article is to describe and analyze the process and evolution of Mexican universities in the world university rankings through a review in a range of the last 20 years. Based on the information obtained from the websites of the rankings of Shangai, QS, THE, Leiden and Scimago added to the elaboration of the state of the art developed with articles on the subject obtained from Web of Science and Scopus, it is possible to describe the evolution of the Mexican universities and understand in depth how they evaluate the rankings analyzed. In the results, a timeline was developed with the evolution of Mexican universities in the main rankings and after explaining the measurement processes, it is presented in a clear and simple way how universities can develop strategies that allow them to reach and improve their position in these rankings	This article aims to delineate and scrutinize the trajectory of Mexican universities in global university rankings over the past two decades. Utilizing data sourced from the rankings websites of Shangai, QS, THE, Leiden, and Scimago, coupled with a comprehensive review of relevant literature from the Web of Science and Scopus, this study offers a detailed examination of the progress of Mexican universities. It provides a deep understanding of the evaluation methodologies employed by the rankings analyzed. The findings include a timeline that charts the progression of Mexican universities across major rankings. Furthermore, the article elucidates the measurement processes and proposes clear and straightforward strategies that universities can implement to enhance their standings in these rankings

References

1. Cid, R., Machin-Mastromatteo, J.D., Tarango, J.: Diagnosing Spanish literature bachelor students' information literacy in digital environments. Digit. Libr. Perspect. **37**(3), 289–304 (2021). https://doi.org/10.1108/DLP-08-2020-0082
2. Frické, M.: XX Conferencia Internacional sobre Bibliotecas Universitarias, en 2023, con sede en UNAM-CDMX [Video]. Facebook (2023). https://www.facebook.com/watch/live/?ref=search&v=656099703277346
3. Ibarra, R., Ballesteros, S.: Infopragmatics: an efficient method for information retrieval: Linguistic solutions for pertinent information. Lambert Academic Publishing. ISBN 978-3843380201 (2010)
4. Ibarra, R.: Linguistic storm: an essential information retrieval tool to update researchers. Qual. Quant. Methods Libr. **4**(1) (2015)
5. Kramer, S.N.: History Begins at Sumer. University of Pennsylvania Press (1956)
6. Roose, K.: Don't ban ChatGPT in schools. Teach with it. The New York Times (2023). https://www.nytimes.com/2023/01/12/technology/chatgpt-schools-teachers.html

Intellectual Property Rights Awareness Among Students at University of Zagreb Faculty of Kinesiology

Ivan Čolakovac and Iva Barković(✉)

Faculty of Kinesiology, University of Zagreb, Zagreb, Croatia
iva.barkovic@kif.unizg.hr

Abstract. In the era of globalization, there is a pressing need to inform college students about intellectual property rights and related law. Real production, the driving force behind every nation's progress, may be created from intellectual property. College students should cultivate their awareness of intellectual property rights because they will be the primary producers and users of intellectual property in the future. Students currently face issues such as lack of education courses, a lack of knowledge reserves, and a lack of legal thinking about intellectual property. Intellectual property is an integral part of the sports world. Sport and its linked activities are developing along with several new phenomena, difficulties, and risks, requiring strict observation and legal compliance. The rise of counterfeiting and piracy has many negative impacts on society and the world of sports. The two primary IP rights in sports historically have been copyright and trademarks. While copyright and trademarks serve as the foundation for many industries (such as broadcasting, fashion, and brand value), it is important to use other types of IP rights to safeguard content and creative work. For instance, the development of advanced technology has led to an increase in research-based, cutting-edge athletic goods. These goods consequently need robust patent protection. Every sport possesses an innovative element, but without IP, sports cannot develop into a profitable asset. In view of these facts, the current study was conducted to determine the level of IP rights (IPR) knowledge among college students at University of Zagreb Faculty of Kinesiology. The descriptive research method was used for this investigation: the study was conducted through a questionnaire. Students of the fourth year of the Integrated undergraduate and graduate university study Kinesiology participated in the research. There were three sections to the questionnaire. The students' demographic data was gathered in the first section. The second half was a list of statements to evaluate students' comprehension of IPR definition and duration of its legal protection, the final section consisted of a list of questions to test students' knowledge in the various categories of IP. The expected results of the survey will show that out of the most known IP rights (patents, designs, copyright, and trademarks) copyright is the most often encountered in students research and therefore the most known to participants. Finally, the outcome of this study will inform projects in the field of Intellectual and Industrial property, such as the ERASMUS + IPEDU project, which aim at forecasting the needs of the future labor market regarding Industrial Property while introducing education and proposing curricula in this field.

A. Katsirikou (Ed.): QQML 2023, 68, pp. 94–104, 2026.
https://doi.org/10.1007/978-3-032-01264-7_8

Keywords: Intellectual property · Higher education · Kinesiology · Croatia · Survey

1 Introduction

Faculty, students, and scientists are involved in research and developing new projects of diverse nature. Many of these activities result in the growth of IPR in the different forms of copyright, patent, trademark, designs, instruments, devices, processes, software, etc. The intricacy of the problem has only risen with the fast expansion of science and technology.

The term *Intellectual Property* is rising more frequently than ever due to the constantly changing technological breakthroughs and thinning international borders. Everyone is vigilant about safeguarding their brand and invention, from information technology to the medicinal and genomic sectors. Understanding intellectual property rights is crucial in this situation.

The definition of intellectual property is "the rights granted to people over the creation of their minds" (Prashant Deshpande et al., 2022). It is a general phrase used in law to refer to patents, copyrights, and trademarks, all of which offer legal protection for ideas, their expression, and the people who came up with them. Intellectual property (IP) is often referred to as intellectual achievement ownership. IP is broadly divided into two categories that are *Industrial Property* and *Copyright* (Prashant Deshpande et al., 2022). Industrial property is inclusive of inventions (patents), trademarks, trade secret, industrial designs, and geographic indications while copyright includes literary and artistic work such as novels, poems and plays, films, musical work, and artistic work.

2 Literature Review

In today's knowledge-based society, IPR are becoming more and more important for sustainable growth. If one has insufficient awareness, it is conceivable to fail to reserve rights for one's intellectual activity. The purpose of study conducted by Prashant Deshpande et al. (2022) was to evaluate the health care professionals' knowledge, understanding, and behaviours regarding intellectual property rights in Belagavi, India. The study found that participants, including academics, have poorer understanding and awareness of IPR. However, faculty members had more knowledge and awareness than postgraduate students did.

Recent trends suggest that undergraduate students are increasingly using works protected by IPR without first getting permission from the owners of such rights. A descriptive survey was created by Igudia and Abioye (2020) to see into student's perception and awareness of IPR. A total of 14.536 undergraduate students from Lead City University and the University of Ibadan in Nigeria make up the population. The approach of stratified and purposeful sampling was used to choose a sample size of 347 students. Data for the study were gathered using a questionnaire, and the Statistical Package for the Social Sciences (SPSS) was used to analyse the data. According to the results, although undergraduate students from the two institutions were aware of the notion of IP, most of them had a poor knowledge of it. Most respondents knew about copyright laws and IP, although a sizable portion did not. The chosen institutions' university libraries did not offer sufficient instruction on copyright problems. Therefore, it was advised that via user education, university libraries should increase students' comprehension of IP and copyright regulations.

Wong et al. (2019) examined Hong Kong's undergraduate students' knowledge of IP. The most well-known IPR, that great number of students were able to accurately identify, was copyright. Further examination found that students from the Business School performed far better than average in providing accurate responses to questions about copyright, patents, and trademarks. This may be because they paid greater attention to IP considering it developed into a valuable corporate asset. Students demonstrated a solid general understanding of copyright protection, according to the findings. Nevertheless, many of them were unaware that copyright does not require registration.

In 2012 Ong et al. also researched undergraduate student awareness at LimKokWing University (LKW) and Multimedia University (MMU) in Cyberjaya, two private universities in Malaysia. They discovered that college students believed plagiarism and piracy were an IPR infringement. Students demonstrated an understanding of how copyrighted materials, including software, music, movies, books and written texts, are violated when they are copied.

Pathan and Anandhalli (2018) researched faculty members' knowledge of IPR at BLDE and SECAB engineering institutions in Vijayapura. The required data was gathered using a structured questionnaire which discovered that most of the faculty members are knowledgeable about IPR rules, regulations, and policies as well as types. Authors conclude that engineers and technical educators must comprehend the importance of intellectual property rights.

The goal of the research project about awareness regarding plagiarism and fair use of copyrighted work conducted by Ahmadi and Sonkar (2015) was to ascertain how well-informed Babasaheb Bhimrao Ambedkar University researchers were on copyright, plagiarism, and fair use. The results of the survey show that most researchers are aware of copyright, although some of them were unaware of copyright violations. It was also noted that the researchers know very little about appropriate usage of copyrighted content. According to the study's findings, many researchers employ free plagiarism-checking programs like Turnitin, Viper, Authenticate, and Plagiarism Checker to prevent fraud.

Popova and Nacka (2017) characterize negative behaviour of the students as their dedication to plagiarism and copyright violations. The goal of their research is to offer advice to non-IPR committed institutions that may exhibit comparable outcomes in terms of student awareness and behaviour toward their academic program and instructional activities. Their research examines whether students with a better degree of awareness and understanding tend to more frequently act positively toward IPR. Not surprisingly, according to the findings, good behaviour is more likely to occur when awareness is higher. On the other hand, while the analysis of student knowledge is related to their degree of awareness, it does not yield any conclusive results. According to the sample, the unfavourable behaviour is caused by the high cost of purchasing the fundamental study materials.

3 Methodology

To get the necessary data from the participants, a questionnaire was created in Microsoft Forms about student's awareness of intellectual property rights. The students' demographic data was gathered in the first section; four questions about sex, age, year of study and average grade. The second part was several statements to evaluate students' comprehension of IPR definitions and duration of its legal protection. The final section consisted of a list of questions to test students' knowledge in the various categories of IPR. The questionnaire consisted of 24 statements concerning patent, trademark, copyright, and design. Students were asked to judge whether a statement is true, false or they were uncertain. Participants in the voluntary survey were made up of fourth year students at the University of Zagreb Faculty of Kinesiology and informed of the study's goal. After quality assessment, 50 students were included in the estimated sample size. Additionally, it was guaranteed that the participants personnel information was kept completely secret, and that the data collected were handled with the highest care. Microsoft Excel was used for data analysis.

4 Results

The research sample's gender distribution, shown in Fig. 1, showed that 34 (68%) of volunteers is male, while the remaining 16 (32%) respondents are women. Questionnaire offered the option to not disclose the gender which no one chose.

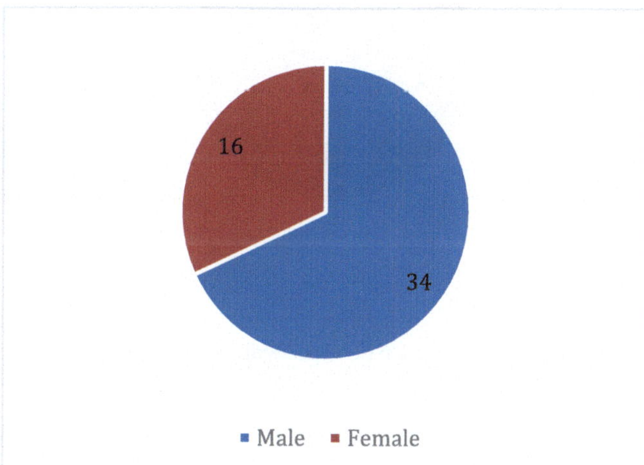

Fig. 1. Distribution of responders by gender

Regarding the age distribution of the kinesiology students, it was found that 27 (54%) participants were 22 years old, nine were 23 years old (18%), seven were 24 years old (14%), and the remaining seven were between 25 and 28 years old (14%) (Fig. 2).

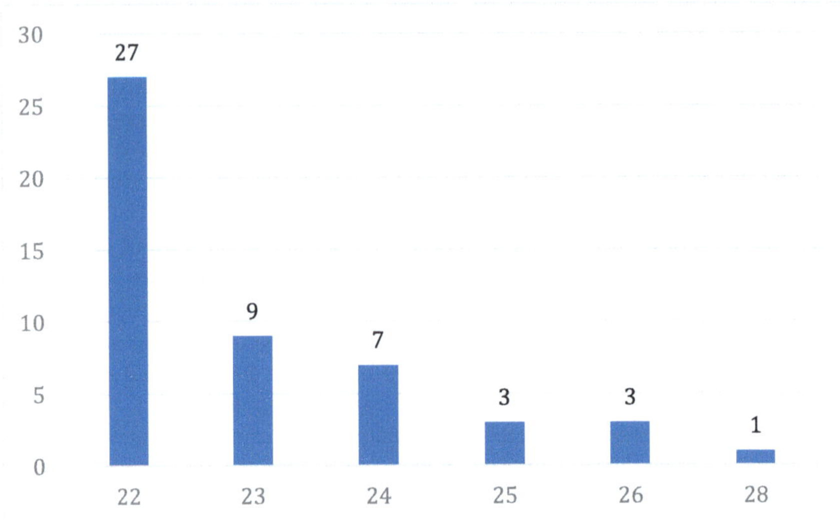

Fig. 2. Distribution of respondents by age

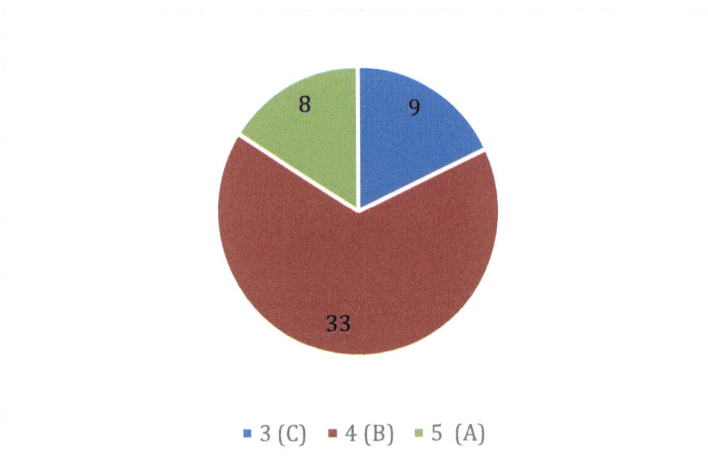

Fig. 3. Distribution of respondent's average grade

As for the average grade, nine respondents (18%) answered 3 (C), 33 (66%) replied 4 (B) and eight (16%) of them have 5 (A) (Fig. 3).

Figure 4 shows student's understanding of basic intellectual property rights: proper understanding of IP definitions was a topic that the students demonstrated a deep comprehension of. Patent was most recognized right (84%), whereas copyright was least understood (75%).

Figure 5 illustrates the level of student understanding of intellectual property rights duration. The respondents' knowledge indicated that they are uninformed on the length

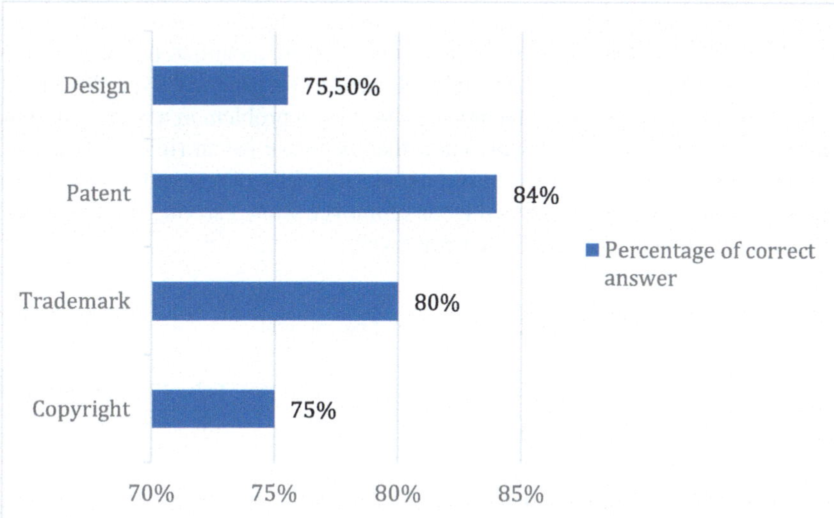

Fig. 4. The percentage of correct responses on connecting intellectual property right with its proper definition

of protection. The results showed that 26% knew how long copyright protection lasts, 20% were informed on the duration of design protection, 12% were familiar with the lasting of trademark while only 8% knew how long patent rights last.

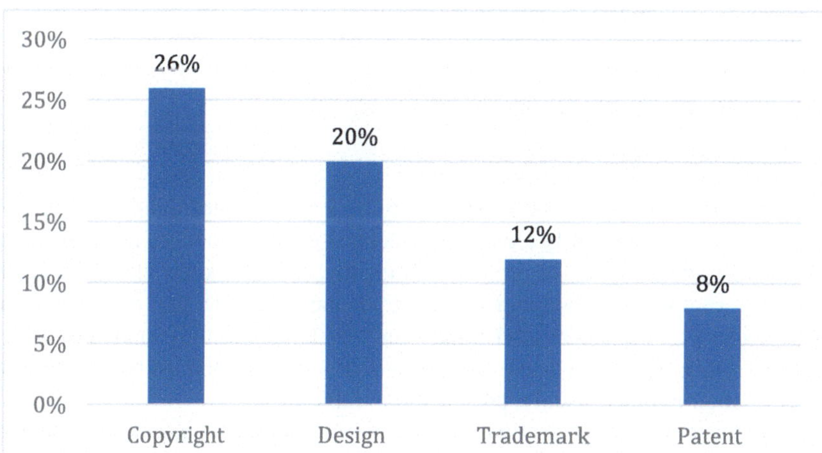

Fig. 5. The percentage of correctly recognized intellectual property right and its duration

According to the statistics, 86% of respondents were aware that patenting an innovation improves a company's reputation and that an invention is considered innovative if it has never been published or exhibited to the public (Table 1). 54% of students knew that disclosing an idea before applying for a patent increases the likelihood that the application would be rejected. Only 30% of respondents correctly identified the statement that inventions do not have to be industrially applicable to be patented, even though 48% knew that a patent must provide a technical solution to a problem in any area of human endeavour. Less than 20% of students knew that exclusive patent rights only apply in the nation or territory where the patent is granted. Only 30% of respondents were aware of the fact that an invention developed by an employee while carrying out the terms of his employment contract belongs to that employee.

Table 1. Knowledge of patent right

Statement	True (%)	False (%)	Uncertain (%)
An invention is considered new if it has not been published or shown to the public before	**86**	12	2
A patent must offer a technical solution to a problem in any field of human endeavour	**48**	30	22
Inventions do not need to be industrially applicable to be patented	48	**30**	22
Telling people about an invention before filing a patent application can lead to an unsuccessful application	**54**	30	16
Exclusive patent rights are applicable only in the country or region where the patent is applied for and granted, in accordance with the law of that country or region	**20**	44	36
If the employee developed an invention in the execution of his employment contract, the invention belongs to the employee	34	**30**	36
Patenting an invention creates a positive company image	**86**	4	10

According to the respondents' knowledge of trademark-related intellectual property rights, 84% of them can identify registered trademark symbols. 64% of participants recognized that a trademark may be protected nationally, and 44% knew that a corporation must specify what it does when registering a trademark. The fact that no one else may register the identical trademark after it has been registered was only known by 12% of respondents. Only 16% of students were aware that a trademark must have a form of a drawing, symbol, or other visual representation (Table 2).

The fact that copyright protection is automatic without the requirement for registration is known to 28% of the respondents. 56% of students were aware that retailers may lawfully play music for consumers without the author's consent, and 58% of students recognized that copyright laws are enforced in the country in which they were created. Although 80% of respondents were aware of the author's right to forbid the use of the work he created, only 50% were aware that the author cannot forbid the changing of the work's content. Student copyright is particularly significant because just 36% of them were aware that faculty cannot publish their theses without the author's consent and that

Table 2. Knowledge of trademark right

Statement	True (%)	False (%)	Uncertain (%)
The symbol® means the name/logo is protected	**84**	6	10
Once a trademark is registered, no one else can register the same trademark	78	**12**	10
For trademark registration, the name must describe what the company does	34	**44**	22
A trademark can be protected on a national level	**64**	4	32
A trademark can only be a drawing, symbol, or figurative form	74	**16**	10

student copyright belongs to them. Only 10% of respondents were aware that copyright still applied to works in the public domain (Table 3).

Table 3. Knowledge of copyright

Statement	True (%)	False (%)	Uncertain (%)
Copyright protection is automatically acquired without registration	**28**	60	12
Copyright laws are enforced in the country in which they were enacted	**58**	12	30
Stores can legally play music to their customers without author permission	24	**56**	20
The author has the right to ban the use of his work	**80**	10	10
The author cannot prohibit changing the content of his work (music, video, etc.)	26	**50**	24
The copyright of thesis belongs to the student, and the faculty cannot publish it without the author's permission	**36**	22	42
Copyright of works in the public domain still lasts	30	**10**	60

Most of the respondents (60%) are aware that industrial design protects an object's ornamental elements. The same number of students were aware that design incorporates two-dimensional elements like lines, patterns, and colour. Only 24% of respondents were aware that an idea cannot be protected by a registered design (Table 4).

Table 4. Knowledge of industrial design

Statement	True (%)	False (%)	Uncertain (%)
Design (industrial) protects the decorative or aesthetic aspect of a certain object	**60**	18	22
Design, apart from three-dimensional form, also includes two-dimensional features such as lines, patterns, and colours	**60**	16	24
A registered design cannot protect the idea	**24**	38	38

5 Discussion

Seven statements made up the group of statements pertaining to patent knowledge. More than 50% of students properly answered only three of the statements. Nearly 90% of students are aware of the issues surrounding the secrecy of inventions. They are aware that they must withhold the invention from the public until it has been patented. 48% of them were aware that a patent needed to address an issue that mankind was confronting and be used in daily life. They had the least understanding of the idea that patents only have exclusive rights in the country in which they were given, as well as the fact that an invention is owned by the employer, not the inventor, if it is made during the employment contract. Balahadia et al. (2022) reach the same conclusion: Philippine students were least aware of the fact that a patent in one country cannot prevent someone from copying an invention in another country.

Only 8% of kinesiology students were aware of the time frame of the patent right. The same applies on the findings of a study by Desphande et al. (2022), which found that just 22% of faculty members and 18.5% of postgraduate students knew that patents enter the public domain after 20 years of registration.

Five suggestions made up the group of knowledge-related assertions concerning trademark rights. More than 60% of students correctly answered the four questions about trademarks while only 30% of students correctly answered the question about trademark registration, demonstrating that respondents' knowledge of intellectual property rights in relation to trademarks is high. Balahadia et al. (2022) came at the same conclusion in their study, which is in line with the respondents' validation that they are aware of trademarks but not their duration.

Data showed that respondents had a basic understanding of copyrights and IPR: nearly 60% of students knew that copyright laws are enforced in the nation in which they are enacted, and 80% of students correctly responded to the statement that the author has the right to forbid the use of his work. Only 28% of them knew that copyright protection is automatically acquired without registration, even though they were aware that stores cannot legally play music without the author's permission and that authors can prevent changes to the content of their works. According to research by Balahadia et al. (2022), students are aware of intellectual property rights thanks to several awareness campaigns. Additionally, respondents claimed that learning about copyright gives them a chance to protect their work that is related to their academic activities.

The respondents are neither aware nor informed, according to the indication for the length of time that copyright is usable. There is no need to register copyright, and if a teacher works full-time at a school, the copyright in the class presentation slides made for instructional use in his or her school typically does not belong to him or her, according to research done by Wong et al. (2019). Students demonstrated low knowledge in their responses to these claims, which match those from our research.

The results of our study agreed with those of Igudia and Abioye's (2020) study on students' perceptions of intellectual property, which was performed at Lead City University and the University of Ibadan. Findings regarding the degree of undergraduate students' knowledge of copyright laws and intellectual property rights showed that a lot of students were knowledgeable about such laws. Results of our research showed that kinesiology students (60% of those surveyed) are aware that industrial designs, which contain two-dimensional elements like lines, patterns, and colours, safeguard the ornamental or aesthetic component of a particular piece. Only 24% knew that an idea cannot be protected by a registered design, similar to research Balahadia et al. (2022) conducted.

After reading the results, it was clear that the most well-known IP types are copyright and industrial design, followed by trademark and patent. This comes as no surprise given the amount of exposure and interaction students have with written works of authors throughout their schooling (elementary and secondary school). The similar finding was drawn by Wong et al. (2019), whose study revealed that 74.3% of students were able to correctly respond to a copyright-related issue. This outcome was anticipated because the Academic Integrity Online Tutorial, which was created to promote academic integrity. Surprisingly, the definition of copyright was the least known among the four IPR, although in higher numbers (75%).

University students can grasp copyrights, patents, trademarks, and industrial designs to some level, according to our study's findings. A better grasp of IPR is essential since the repercussions for infringement of the protected works, ideas, and creations might result in sanctions. The responses attest to a lack of education when it comes to the length of time that a patent, trademark, copyright, and industrial design remain protected.

6 Conclusion

The ever-evolving field of research is growing in popularity across all industries which demands awarness of intellectual property. To protect a creator's ideas, one must have a fundamental understanding of their legal rights. The results of this study show that students require some thorough explanations of IPR and its application. The findings, notwithstanding the limited sample size in one faculty, may only be viewed as preliminary and offer some guidance for further research. To verify the results, similar studies might be conducted at other universities and institutions of higher learning. IP education must be incorporated into university learning programs due to the growing significance and value of intellectual property rights. It is advised that the institutions create resources, like a guidebook, on intellectual property rights for wider distribution for students.

Additionally, continuing to hold seminars, projects and training sessions will help both the faculty and students. To prevent any misuse, students should be given examples of the meaning, function, and differences of IPR.

References

Ahmadi, A., Sonkar, S.K.: Awareness regarding plagiarism and fair use of copyrighted work: a survey amongst Doctoral Students of Babasaheb Bhimrao Ambedkar University, Lucknow. J. Inf. Manage. **2**(2), 98–110 (2015)

Igudia, O.E., Abioye, O.A.: Perception and awareness of intellectual property rights by undergraduates of two universities in South West, Nigeria. Global Sci. J. **8**(7), 1596–1614 (2020)

Ong, H.-B., Yoong, Y.-J., Sivasubramaniam, B.: Intellectual property rights (IPR) awareness among undergraduate students. Corp. Ownersh. Control.Ownersh. Control. **10**(1), 711–714 (2012)

Pathan, F., Anandhalli, G.: Awareness of the intellectual property right among the faculty members of BLDE and SECAB engineering college Vijayapura (District) Karnataka. Int. J. Sci. Res. **8**(3), 1107–1113 (2018)

Popova, K., Nacka, M.: Intellectual property rights knowledge and awareness: academic level empirical analysis and recommendations. J. Contemp. Econ. Bus. Issues **4**(1), 41–54 (2017)

Prashant Deshpande, A., et al.: Assessment of knowledge and awareness regarding intellectual property rights among the health-care professionals in Belagavi city: a cross-sectional study. J. Educ. Health Promot.Promot. **30**(11), 211 (2022). https://doi.org/10.4103/jehp.jehp_967_21

Wong, K.C.K., Lee, F.S., Choy, M.M.C.: Intellectual property: how well undergraduate students aware of? Int. J. Knowl. Eng.Knowl. Eng. **5**(2), 72–75 (2019)

Information Services

Žiča's Poetry Festival *Transfiguration*

Ana Gvozdenović(✉) and Mirjana Savić

National Museum Kraljevo, Kraljevo, Serbia
gvozdenaana@gmail.com

Abstract. For more than 30 years, Public Library and National Museum in Kraljevo, Serbia, have organized cultural manifestation Transfiguration. The idea was born in 1990's with a movement that included revitalising of Serbian medieval heritage. Only 5 km from the city center there is Žiča Monastery, built in 1206–1221, the place where the first Serbian kings were crowned. Nowadays, we are crowning poets here and the final of the great poetry festival takes place in this monastery. The festival is part of the Versopolis project from 2018 and it shows how a new tendency in cultural promotion can be united with traditional values. During the festival, the Library and Museum, with the Literary Club (civil society) organize many different activities: literary evenings, critical meetings, conferences (with eminent historians, archeologists, archivists…). We published many papers dedicated to the cultural heritage of our city and, on the other side, to contemporary Serbian poetry. In the last few years we brought to Kraljevo regional and international authors too and we prepared some programs for young and Gymnasium students. In this way, we successfully connect tradition and modernity, cultural heritage and contemporary poetry.

Keywords: Poetry · Cultural heritage · Awards · Žiča monastery · Public library stefan prvovenčani · National Museum Kraljevo

The Spiritual Congregation of Žiča "Transfiguration" has been taking place for more than three decades as a cultural and academic event founded with a view to nurturing, researching and affirming the authentic values of the national culture. After the break-up of the Socialist Federal Republic of Yugoslavia in the civil war that started in 1991, and the creation of a curtailed Federal Republic of Yugoslavia in 1992, with inflation, economic embargo and the growing economic crisis as well as waves of uprooted Serb refugees, a need emerged in Kraljevo for the city to grow beyond that cruel time and to transform the society through the messianic power of faith. In 1992, by the initiative of the Literary Club of Kraljevo, "The Days of Transfiguration" in Žiča were organised for the first time. The place and the date were selected as a symbolic representation of the very essence of the event: on the one side, the foundational tradition of the Serb spirituality and statehood, as epitomised by the Žiča Monastery; on the other side, the religious feast day Transfiguration on 19 August, as a symbol of spiritual, cultural and intellectual transformation of Serbia. The festivity started with the endowment of 1500 books to the Žiča branch of the Library of Kraljevo. This was followed by a book promotion event for five authors from Kraljevo whose books were published in the series

A. Katsirikou (Ed.): QQML 2023, 68, pp. 107–114, 2026.
https://doi.org/10.1007/978-3-032-01264-7_9

"Slovo" with the Literary Club Kraljevo and the Graphic-Publishing house "Slovo." It was decided that each year a literary author be awarded a *Žička hrisovulja* (*Žiča Charter*) award and an original icon. The first to receive the award was the poet Dobrica Erić. The performance of the octet of the choir "Saint Sava" of the Žiča Eparchy as well as poetic and acting interpretations of poets and theatrical artists accompanied the event, giving it a solemn tone. In addition to the Literary Club of Kraljevo, other cultural institutions of Kraljevo teamed up to organise the next event in 1993. The sponsorship of the festivity was taken over by the Municipal council of Kraljevo (as of 2008, the City of Kraljevo). The expert supervision and the organisational tasks were taken over by the Library of Kraljevo, the Žiča Monastery, the National Museum of Kraljevo, the Theatre of Kraljevo, the Institute for Protection of Cultural Monuments of Kraljevo and other cultural institutions in accordance with the programme. The Žička hrisovulja was defined as an award "for poetic works which bring together traditional and contemporary currents of Serbian poetry". It was decided that the award jury for the Žička hrisovulja would consist of five members who on the moveable feast day of Ascension would announce the name of the poet laureate for the given year. The programme of the Assembly begins with a ceremonial opening on 16 August (except in 1995 and 1997 when the event started earlier), with selected artistic and musical acts. The main part of the festivity is dedicated to literature with a special emphasis on the creative work of the poet laureate, with exhibitions, scholarly events, musical events, book promotions, theatrical plays, orations, and lectures. The programme finishes on 19 August with the award ceremony for the poet laureate in the dining chamber of the Žiča Monastery. In the early years of the festival, the poets used to receive the Žička hrisovulja produced in the monastery icon painting workshop "Saint Luke": the icon was painted on lime wood with a golden background and combined iconography and text. From 1999 onwards, in cooperation with the Kraljevo artist Slobodan Luković, as a special tribute to the poet laureate, the icon of Transfiguration is awarded along with the Žička hrisovulja. The hrisovuljas are produced on parchment with a caligraphic text, miniatures and wax stamp. The spiritual aspect of the festivity has been confirmed by the blessing of the Žiča eparchs Stefan, Hrizostom, and Justin. (Gvozdenović and Savić 2011).

Over 32 years of its existence, the festival changed its name three times. In the first year, it was named "The Days of Transfiguration" then in the period 1993–1996 it was known as the Serbian religious congregation "The Days of Transfiguration" and from 1997 onwards it gained the name the Žiča Spiritual Congregation "Transfiguration" which remains unto the present day. Crucial to the organisation of the festivity is the financial support of the City of Kraljevo and the Ministry of Culture of the Republic of Serbia; in 1995, the Ministry of Science and Technology was also involved by supporting an academic conference as part of the Congregation. From the very first days of the festivity, a group of people from the Library of Kraljevo, the Literary Club of Kraljevo and the National Museum of Kraljevo distinguished themselves through expertise and personal efforts that contributed in a significant way to the delivery and promotion of the event. In lasting, and evolving, over years, the festivity retained the core elements of its 1992 programme with a sophisticated and multi-layered scholarly development. The power of words and verses gained worthy followers in the sundry features from the humanities and different arts which through their themes and their significancy largely surpassed the regional format.

As part of the Žiča Spiritual Congregation, from 1994 to 2022, the National Museum of Kraljevo organized 39 exhibitions, eight book and magazine promotions, six expert lectures in the fields of history, art, ethnology and conservation, a series of concerts, several sermons and theater performances and, what is particularly significant, six scientific meetings, round tables and symposiums: *Cultural heritage and its protection* (1994), *Žiča-Art, History* (1995), *Rudo Polje-Karanovac-Kraljevo: from the first mention to the First World War* (1997), *Kraljevo–Rankovićevo-Kraljevo* (2015), *Journalism in the Development of the State and Society–from Press Reports to Objective History. Karanovac–Kraljevo in Serbian Press 1835–1918* (I, 2021), *Journalism in the Development of the State and Society* (II, 2022). Thereby, this festivity became a place where the Serbian cultural patrimony is keenly preserved, carefully nourished, and thoughtfully developed: the entire cycle, from the initial idea to a fully fledged national programme, took shape under its aegis. In 1995, a scholarly conference was organised under the title *Žiča-Art, History* and chaired by four academicians (Pavle Ivić, Dejan Medaković, Sima Ćirković and Vojislav J. Đurić). The objective of this gathering was for the multi-layered whole, formed by the Žiča monastery and the Žiča Archeparchy to be approached through a comprehensive involvement of 45 most eminent scholars. Along with this scholarly event, an exhibition *entitled Research and Renewal of the Žiča Monastery* was organised under the curatorship of Milka Čanak Medić. A bibliographic catalogue of the Žiča monastery was published in 1998. Two years later, the edited volume *The Žiča Monastery* came out of press and the Žiča Spiritual Congregation of 2001 was an occasion for these significant books to be officially launched. A similar trajectory characterises subsequent scholarly events which have been followed by valuable edited volumes as an indispensable starting point in nearly all research in the field of humanities.

The book launch of two out of a total of four books about Karanovac–Kraljevo in Serbian press from 1835 to 1918, edited by Slavoljub Maržić in the framework of the Žiča Spiritual Congregation, disseminated to the general public a part of this author's research which enriched the national historiography and made the Kraljevo bibliography globally unique. These volumes inspired and set a firm foundation for organising three scholarly conferences with international guests, "Journalism in the Development of the State and Society" (2021, 2022, 2023) in cooperation between the Natonal Museum of Kraljevo with the Faculty of Law in Belgrade and the Institute for Political Studies in Belgrade.

A particular attention was paid to the local museum exhibitions in Kraljevo as well as visiting exhibitions which were in harmony with the values and the aspiration of this festival to be open to new visions and themes while remaining traditional in its foundation. Often, two exhibitions were organised within the same annual gathering in order to make its features richer, its varied information more accessible and its message as comprehensive as possible. Segments of our nation's heritage were on display at the following exhibitions: *Tombstones from the Collection of the National Museum of Kraljevo* and *Rudo Polje–Karanovac–Kraljevo* (1997); *Formal Attire in Serbia in the 19th and the 20th Centuries*, The Historical Museum of Serbia (2003), *Crafted Weapons of the Balkans 17–19th Centuries*, The Military Museum of Belgrade (2004),

The *First Serbian Uprising*, The Historical Museum of Serbia and the National Museum of Kraljevo (2005), *Libade*, The National Museum of Kraljevo (2009), *Whisper Through Eternity–Carpets from the Ethographic Collection of The National Museum of Kraljevo*, the National Museum of Kraljevo (2017), *Under the Golden and Blue Sky–the Icons of Greek Masters from the 15 th to the 19 th Centuries in the National Museum of Serbia* (2019). Contemporary currents or artistic creativity were represented by the exhibitions *Dragan Pešić–Graphic Design* (2004), *Slobodan Štetić–Poster*, The National Museum of Kraljevo (2015), *Museum Stories–Acquisitions 2014–2017* (2018) and others.

As part of the "Transfiguration" festivity, further two exhibitions have been organised in cooperation with foreign embassies: *The Colours of Orthodoxy–Poland*, the exhibition of the Polish Embassy in Belgrade and the Ortphoto portal (2014) and the exhibition of the Serbian artist living in Canada *Bratsa Bonifacho*, supported by the Canadian Embassy in Belgrade (2020).

A continuity in presenting the most significant historical landmarks, but also in recognising historically important moments in the present day, has been achieved with the exhibition *The Eight Centuries of Chilandar* (1998) and then again with the exhibition *The Reconstruction of Chilandar after Fire* (2010).

The exhibition *The Time of Light–The Serbian Art of the 13 th Century* (2007), curated by Suzana Novčić and Tatjana Mihajlović, was organised to mark the eight centuries of the Žiča Monastery and staged in 14 cities with acclaim, becoming the first exhibition of the National Museum of Kraljevo to be presented in the SANU Gallery in Belgrade. An exhibition marking two decades of the Congregation, *Transfiguration, 20 years* (2011), curated by Ana Gvozdenović amd Mirjana Savić, was initially presented in The National Museum of Kraljevo and then at the Book Fair and the Student Cultural Centre in Belgrade. This demonstrated that the fruits of the cooperation of the Library and the Museum have become a recognisable and independent ensemble, consisting of selected and diverse features. A decade later, Ana Gvozdenović and Mirjana Savić went on to curate the thematically expanded and visually upgraded exhibition *Transfiguration, 30 Years* (2021), which was shown as a guest exhibition in Niš, as part of the festival The Poets' Colony of Sićevac.

Chiming in with the core contents of the programme, promotion events were organised for scholarly publications and the journal *Our Past*, published by the National Museum of Kraljevo and the Historical Archive of Kraljevo. Likewise, a number of lectures of eminents scholars have been collated and published: those of Slavenko Terzić about the Old Serbia, of Aleksandar Dodig about caligraphy, of Olivera Gavrić about the mosaic art in Serbia 1950–2015, and of Aleksandra Novakov about Serbian education in the Ottoman Empire 1878–1912, etc. Within the Žiča Spiritual Congregation "Transfiguration", appropriate programmes have been designed to mark sgnificant anniversaries and events. The archpriest sermon of Stefan, the Eparch of Žiča, in the Orthodox Spiritual Centre "Vladika Nikolaj Velimirović" on the occasion of the most significant anniversary for all Christians–2000 years of Christianity–attracted the attention of a large audience. The 150th anniversary of the institutional protection of the cultural heritage in Serbia was marked with a roundtable *Cultural Legacy and Its Protection* (1994). The eight centuries of the Chilandar Monastery was an occasion for the eponymous exhibition and the suitable musical programme as part of the 7th Žiča Spiritual Congregation "Transfiguration" in 1998. The eight centuries of the existence of the Žiča Monastery inspired the National Museum of Kraljevo to prepare a remarkable exhibition *The Time of Light–the Serbian Art of the 13 th Century*. On the occasion of the 17 centuries of the Edict of Milan, which put an end to the persecutions of Christians and declared equal rights to their religion, an exhibition of Veljko Mihailović's graphic art, entitled *Constantine*, was curated in 2013. The centenary of the begininng of the First World War in the museum of Kraljevo provided an occasion for a series of exhibtions and programmes, including the guest exhibition *Serbia and Serbs in the First World War in Films from the Collection of the Yugoslav Film Archive* (2014). The exhibition *Our fellow citizen Queen Draga–Between Glory and Anathema*, staged in 2021, marked 120 years since the queen's last visit to Kraljevo, where she had lived from the age of 16 to 19. Sixty years since the change of the city's name from Rankovićevo to Kraljevo inspired The National Musem of Kraljevo to organise the scholarly conference *Kraljevo–Rankovićevo–Kraljevo* (2015). Over the three decades of its existence the Žiča Spiritual Congregation "Transfiguration" succeeded in exceeding even the most ambitious hopes and the most optimistic expectations of the members of the Literary Club of Kraljevo who first convened in Žiča in 1992. Initially launched as a purely literary festival, the wealth of the Congregation's themes soon attracted a wider circle of readers, artists, scholars, and of all those who saw spirituality as an existential necessity and poetic verse as a symbol of collective and intimate endurance. From year to year, the Congregation created a series of assorted artistic features and expert programmes which rely on scholarly methodology to shed light, or otherwise open a new perspective, on the national history and creativity. The Žiča Spiritual Congregation "Transfiguration" is now considered as one of the most important cultural events in Serbia.

Over a period of more than thirty years, the Žiča Spiritual Congregation asserted itself by concise and clearly defined features. The backbone of the event was the poetic prize the *Žička hrisovulja* awarded to poets whose overall opus reaches a high standard within the contemporary Serbian literature, one that grows out of the foundational currents of the Serbian poetic tradition and that fuses this tradition with the poetic currents of its own time (The Statute of the Congregation). In the multitude of literary awards in Serbia, the *Žička hrisovulja* is recognizable through its symbolic value as well as through dedicated edited volumes about its poets laureates, which remain a memento of the event and a lasting contribution to the research of the contemporary Serbian poetry. Some other literary panels have followed the lead of Kraljevo and paired their awards with edited volumes about the work of the award holders.

The laureates have been the most important Serbian poets whose poetry rests on the Orthodox spirituality of Žiča but which also expands the range of contemporary Serbian poetry in a significant way: Miodrag Pavlović, Ljubomir Simović, Ivan V. Lalić, Matija Bećković, Milovan Danojlić… Shortly before our departure to this conference, the award jury has announced that the laureate for 2023 is the poet for children and adults Ljubivoje Ršumović, "for poetry which many generations of our child-like nation have grown up with". Perhaps the laureates of Žiča are best described by Ivan V. Lalić's imperative to say *yes* to this world despite the seductive notion of its senselessness. Each awarded author has lent credibility to the award by the uniqueness of his own poetic expression and also recommended the subsequent laureates to the general readership and the scholarly audience.

Unsurprisingly, the largest part of the congregation's features is dedicated to the laureate of the year. He is honoured by several topical programmes in the city. The first literary event of the Congregation is *The Evening with the Laureate*, where the poet himself reads his poetry and thereby introduces his work to the general audience. Regardless of whether it happens in the packed main stage of the theatre or in the more intimate atmosphere of the studio stage this is always a festive occasion where the audience can listen to the live reading of the awarded author.

On the eve of the feast day of Transfiguration, the laureate's poetry is taken up by literary experts, researchers and interpreters. In the Reading room of the Library of Kraljevo, a scholarly conference is organised to debate the laureate's work. The organizer of the conference also acts as the editor of the volume with conference proceedings. In the past, the conference has been chaired by some of our most eminent scholars of the contemporary Serbian poetry, including academicians and university professors, and its participants included historians and theorists of literature, linguists, literary critics as well as experts from other fields (anthropologists, psychologists…). The conference organizer has a clear goal to shed light on the creative work of the laureate from multiple angles and to enable new plausible readings.

The laureate's work is also the theme of an exhibition which is staged in one of the art galleries of Kraljevo and which is subsequently hosted in Belgrade and some other cities of Serbia.

The central event of the Congregation is the ceremonial award of the *Žička hrisovulja* to the laureate on the feast day of the Transfiguration of the Lord, 19 August, in the dining chamber of the Žiča Monastery. The laureate receives the award from the president of the Organising committee and his poetry is acknowledged by researchers of his work and the president of the jury panel. The awarded poet then addresses the gathering with a speech which may be socially engaged or otherwise intrinsically poetic yet always distinctive and valuable: a gift to numerous visitors gathered in the Žiča Monastery on the feast day as well as a gift to the entire Serbian literature. It is precisely for this reason that the Library of Kraljevo marked the 20th anniversary of the Congregation by publishing a collection of all speeches of the laureates up to that point.

There is a special interaction between this festival and the publishing activity of our Library. Not only are our editors members of the award jury, our Library has also published 27 edited volumes (with a 28th in the pipeline) dedicated to the poetry of contemporary Serbian poets. These edited volumes make a relevant reading list for students of Serbian literature but also an indispensable material in any serious study of Serbian poetry.

We would like to finish this paper by quoting the words of Mihajlo Pantić, a writer and professor at the Faculty of Philology in Belgrade, addressed on the occasion of the 20th anniversary of the Congregation:

„Poetry is the foundation of any culture. This fact is confirmed by man's ancient need to turn into verses whatever needs to be said about life, whatever there is in human experience, but also, to no less extent, in the individual and collective imagination; yet, this fact is accidentally or deliberately forgotten in modern times and rarely recalled.

There is no European culture without Homer, no English culture without Shakespeare, no Russian culture without Pushkin, no French culture without Baudelaire, no Polish culture without Mickiewicz and this series can go on into infinity but the conclusion is unavoidable, inevitable and ineluctably true: where there is no poetry there is no culture in any sense of this concept. Thereby by poetry I understand not only what is said in verse but also the aestheticised essence of every culture, regardless of the medium which it uses to shape and express itself. Poetry is, therefore, the salt of the salt, the core of the core of humans' individual and collective existence, it is equally important to the idiorhythmics and to communal life; without poetry's capacity to transcend our life, there would be less meaning to the world and fewer reasons for us to exist, if that meaning and reasons existed in the first place.

And it is precisely for this reason, through inert forces of our being which nowawdays, as well as in all historical epoques, is exposed to different temptations, that we are prone to forget the important, more precisely categorical, facts and to glorify triviality, petty interests, and inflammable yet easily consumed passions. It is necessary constantly to repeat, in all possible, new and old, versions and variants, that same thought which the European culture has been reiterating ever since it first occurred to Plato: the virtues of the state reflect those of its music, and poetry is the music of language, and the music of the human soul."

On the same occasion Professor Pantić called Kraljevo the decades-long capital of Serbian poetry, "specifically that kind of poetry which emphasises the human need to turn its face towards the divine light": "Kraljevo is a city that has not forgotten its 30-year old custom—and customs do oblige—for poets to get together for the feast day of Transfiguration to read and celebrate poetry and for one of them to continue the sequence of those who can be said to represent a thoroughfare of contemporary Serbian poetry. What this thoroughfare does is to contribute into the treasure-chest of the Serbian poetic discourse its very best an highest gifts. The days of Transfiguration, which culminate in the award of the *Žička hrisovulja*, bring closer and fuse two of the greatest traditions of ours, namely the sacral and the poetic one. They are a festivity of the first order of magnitude and ought to be approached with the due respect and with a certitude that they are going to last even when we pass, when new epoques bring new poets and interpreters who will know what their most important task and duty are: to continue and contribute in their own way to the same tradition we are nurturing in the present day. This sequence is simultaneously a plausible and highly legitimate anthology of the Serbian poetry of the past decades." (Pantić 2011).

Hence, for more than three decades, and through the efforts of multiple cultural agents, the Kraljevo festival has been merging spirituality and art, building on the foundations of our national being, yet never losing its connection with modernity.

Following the contemporary trends, for a few years already, our festival has been included in the platform of the European poetic festivals Versopolis, initially as its associate member and then as a full member.

VERSOPOLIS is a platform of major European poetry festivals which, through mutual collaboration, connects them into a wider pan-European program intended for: discovering and encouraging mobility and pan-European promotion of emerging poets from different European countries; thinking of new ways to involve existing and new audiences in the creative process; and creating an unified European poetry field that will allow quality artists from different countries to establish themselves transnationally.

VERSOPOLIS has created specific activities which enable the development of European poetry, with the intent of creating an added value for the cultural and creative sector based on developing and utilizing common creative and innovative operating methods and with the intent to create a new European festival perspective under a joint brand of European poetry festivals–Versopolis.

The main objectives of VERSOPOLIS include:

„Establishing emerging European poets from different countries in the wider European area through increased mobility of the poets and their works.

- Developing poetry festivals and creating a new perspective of poetry festivals for long-term collaboration in the future.
- Creating a joint brand of cooperating European poetry festivals and increasing the recognisability of highquality European works within the wider European space." (https://www.versopolis.com/)

Owing to this project we have been able to host poets from Macedonia, Slovenia, Croatia and this year we hosted authors from Hungary and Greece.

Finally, we would like to invite you to our gathering; we are sure that you will get to know the Žiča Monastery, the source of the Serbian spirituality and statehood, and also that you will find reflections of your own scholarly or artistic interests.

Literature

The Statute of the Congregation. https://www.versopolis.com/

Гвозденовић, А., Савић, М.: *Преображења: 30 година Жичког духовног сабора* (стр. 46). Народни музеј Краљево (2021)

Gvozdenović, A., Savić, M.: Žički duhovni sabor Preobraženje – mesto gde je baština živa i gde se čuva, u *Libraries, Archives and Museum Conference (LAM)*, Montenegro, Zbrnik radova, Prva međunarodna konferencija bibliotekara, arhivista i muzeologa, (pp. 179–189) (2018)

Гвозденовић, А., Савић, М.: *Преображења: 20 година Жичког духовног сабора* (стр. 36). Народни музеј Краљево (2011)

Пантић, М.: Прошло је двадесет година. Повеља **41**(3), 184–185 (2011). https://povelja.kv-biblio.org.rs/wp-content/uploads/2021/03/2011-br.-3-Proslo-je-dvadeset-godina-str.-184-185.pdf

Песници у Жичи: беседе на Преображење. (ed.): Драган Хамовић. Краљево. Народна библиотека Стефан Првовенчани (2011)

Ethical Information Factors for Shaping Value-Sensitive Information Services of Libraries

Jela Steinerová(✉)

Faculty of Arts, Department of LIS, Comenius University in Bratislava, Bratislava, Slovak Republic
jela.steinerova@uniba.sk

Abstract. The paper is aimed at determination of ethical factors which can shape value-sensitive information services of academic libraries. It is based on analyses of selected theoretical frameworks of information ethics studies and on an own empirical Delphi study of experts in the Slovak and Czech Republic. In the analyses we applied discourse analyses, content analyses and conceptual modelling. We present a resulting conceptual model of ethical factors of human information interactions. We proposed the inclusion of the identified ethical factors into models of information literacy and human information behaviour. In conclusions, we introduce a concept of value-sensitive information services of academic libraries focused on verification of information.

Keywords: Information ethics · Information behaviour · Information literacy · Mis/disinformation · Model of ethical factors · Value-sensitive services · Academic libraries

1 Introduction

The purpose of this paper is to determine ethical factors of information work and include these ethical information factors into new models of value-sensitive information services of academic libraries. Our aim is to find understanding of selected ethical challenges of information use in digital information environment. We ask the question: Which ethical factors shape the use of information and visions of new services of academic libraries?

Information ethics is a multidisciplinary field of study which is focused on applied values of information, ethical perspectives of access to digital information, intellectual property, privacy and personal data, accuracy of information and online sharing and communication of information. The ethical perspective is both descriptive focused mainly on understanding factors, rules, norms, virtues and values of information, and normative focused mainly on norms and rules connected with information behavior, information literacy and moral literacy.

In this paper we focus on explanation of ethical challenges of digital information work in the contexts of ethically informed human information interactions, including ethically informed use and ethically informed production of digital information. We suppose that

A. Katsirikou (Ed.): QQML 2023, 68, pp. 115–124, 2026.
https://doi.org/10.1007/978-3-032-01264-7_10

ethical framework of information experience can help conceptualize value-added and value-sensitive services of academic libraries.

The first part of the paper reports on the analyses of selected theoretical framework with regard to ethical information factors and ethical dilemmas of human information behavior and information literacy. In the second part we use the data from a Delphi study of experts focused on ethical challenges of digital information in academic libraries. In conclusions we present a model of ethical factors of human information interactions and proposals for value-sensitive services of academic libraries.

2 Theoretical Framework and Methodology

The theoretical background of the study was based on analyses of selected works focused on information ethics (Floridi 2013, Kelly and Bielby 2016, Capurro and Pingel 2002, Steinerová 2020, 2021). We also used bibliometric analyses of published works on information ethics registered in WoS (461 records) and SCOPUS (653 records) databases in the years 1988–2020. These analyses confirmed main topics of information ethics focused on access to digital information, privacy and security, accuracy of information (disinformation, truth etc.), intellectual property and ethics of online communication and communities (Burgess and Knox 2016, Bawden and Robinson 2020, 2022). Emerging topics include mainly ethics of artificial intelligence, data ethics and data literacy, ethics of information literacy, issues of dis/misinformation and ethics of social media. In the academic environment the dominant topics are academic integrity, research integrity and ethics of open access and digital publishing (Capurro 2019, Floridi et al. 2018, Floridi 2019).

In detailed content analyses we have reviewed selected models of human information behavior and information literacy (Bruce 2016) with regard to dis/misinformation. Disinformation is explained as purposefully untrue or incorrect information in the contexts of social perception of information (Karlova and Fisher 2014, Walton et al. 2021, Lewandowsky et al. 2017), social diffusion of information (Ruokolainen and Widén 2020) and causal factors as context (Agarwal and Alsaeedi 2021, Agarwal 2022). Selected models of information literacy were considered which explore and include information ethics, namely metaliteracy (MacKey and Jacobson 2019), moral literacy (Tuana 2007), the framework ACRL (2016), qualitative landscape of information literacy (Lloyd 2021) and the model ANCIL (Secker and Coonan 2013). These models consider ethical awareness, ethical sensitivity, values of information and moral imagination as part of ethical information use, sharing and production (Steinerová 2022a, b, c). We have used the concept of human information interactions (HII) as determined by Fidel (2012) and Marchionini (2008) and information ecologies (Nardi and O'Day 1999, Davenport 2018) as a context for searching for ethical factors of human information behaviour and information literacy focused on digital environment. Another applied context is represented by the academic environment, especially values of academic libraries (Frandsen and Sørensen 2020, Gorman 2000, Oakleaf 2010, Saracevic and Kantor 1997, Urquhart 2018).

Based on the analyses, main research questions of the study were articulated as follows: Which ethical factors are most significant with regard to information use and production? How can we include these factors into conceptual models of information behavior and information literacy? Can these ethical factors shape new value-added services of academic libraries?

3 Results of a Delphi Study: Content and Discourse Analyses

As part of a wider research project, we have designed a Delphi study on challenges of information ethics in digital environment. The study was divided into a pilot study (4 experts), main study/survey of 19 experts and an online discussion of 6 experts. The experts were selected from academia and practice of libraries and ICT companies from Slovakia and the Czech Republic. The main disciplines included information science, computer science, media, marketing, management, psychology, political science, but also experts from academic libraries. The purpose of the Delphi study was to find out what is the perception of ethical information issues by experts and how they assess potential future. The main question is what are the main ethical challenges of information use in digital age. The design of the Delphi study was composed of three main questions, namely the perception of three main ethical issues of digital information, three main ethical challenges of ICT and AI, and three main values of information. The acquired data was analysed independently by two researchers in separate phases, which included data from a pilot study with 4 experts (December 2021), the main study—an online survey in January–May 2022 (19 experts, 13 from SR, 6 from CR) and an online discussion with 6 experts (September 2022). In the analyses we applied content analyses, qualitative analysis of data, discourse analysis and conceptual modelling, managed by two researchers. Resulting categories were discussed and reviewed by selected participants in order to reach validity of results and balance the limits of subjectivity. As for gender, 19 participants included 12 men and 6 women, online discussion 3 men and 3 women. As for disciplines, 5 experts came from library and information science, 5 experts from informatics (including social informatics) and 1 expert from psychology, management, philosophy, political science, 4 experts from academic libraries, 1 expert from IT company. A majority of participants came from academic institutions (13 / 68%), the rest were managers (5).

Results of the study point to the epistemic consensus of experts with regard to three questions articulated as part of the main study. These results were visualized in several conceptual models. We identified three main ethical issues of digital information work, namely privacy and personal data, benefits and risks of intelligent technologies, and disinformation. Several ethical challenges were mentioned, especially balance between public and private, social benefits and social risks of digital technologies and artificial intelligence or decline of ethical frameworks in society and ethical crisis of information work. Main ethical dilemmas included the impact of technologies on human information interactions, bias of AI and algorithms, and risks of social control and management (e.g. social power structures, limits of democracy). As for the use of AI, main ethical risks were mentioned as value tensions embedded in the use of advanced technologies for information use, sharing and production. Three main values of information were identified by participants as utility of information, truthfulness and objectivity / credibility of information. Further related categories emphasized critical thinking, human responsibility and information literacy. Relevance assessment was determined as part of social and epistemic utility of information. The abilities to recognize the "toxic" information and verify information were related to responsibility and trust. Some participants mentioned the abilities related to sense-making and making meaning of information with regard to context. Discernment of true information were mentioned in the context of pollution of digital social media with dis/misinformation. Knowledge from previous studies was confirmed which point to social perception, social diffusion of information, cognitive bias, emotions and motivated reasoning as causes of dis/misinformation. In concluding synthesized models, we have identified social representations of ethical challenges of digital information use, sharing and production. Main ethical components and factors are social, cultural and intercultural rules, value tensions between people and technologies, and the truth and utility of information in the application areas of education, research, workplace and everyday human information interactions.

These results were compared with results of prior bibliometric analyses. Trends of topics related to information ethics research identified such main topics as ethics of artificial intelligence, data ethics, research and publishing ethics, academic integrity, ethics of information literacy and dis/misinformation. Further related topics include digital culture, metaliteracy, ethics of online searching and information sharing in social media.

The discourse analysis of the online discussion revealed four main categories related to the question of future development of information ethics studies. These include the determination of an ethical crisis in the society, need for ethical education of top managers and designers of AI, ethical issues of AI systems, human background of ethical information interactions. With regard to the impact of technologies on ethical factors of information use, the attitudes were divided into the optimistic, pessimistic and realistic. Main possible tools for further development of information ethics in future were determined, such as education, information literacy and collaboration of people and AI tools.

The study has found that main ethical factors which should be included in models of human information behaviour, information literacy and information interactions are social and intercultural rules, value tensions and co-evolution of people and technologies with focus on discernment of truth and utility of information. The experts identified the need to strengthen ethical sensitivity and ethical awareness of society with regard to information use and production. Potential means for dealing with ethical crisis of digital information are mainly collaboration of different disciplines and experts, common concepts and procedures, especially in development of academic education, academic integrity and information literacy.

4 A Conceptual Model of Ethical Factors of Human Information Interactions

Based on the results of the Delphi study we have proposed a conceptual model of ethical factors of human information interactions (Fig. 1). It is composed of interconnected strata of ethical factors of information use, as identified in the study and in the theoretical analyses. These interconnected components include selected ethical factors which can play an important part in information use and production in digital environment, namely ethical sensitivity and awareness (as part of information literacy), ethical imagination and intuitions (related to emotions), ethical decision-making and problem-solving related to knowledge of tools, norms and rules in information use, sharing and production, and ethical metacognition, including mainly self-regulation in the information environment. The contexts of the model are social, cultural and historical factors, assessment of truth of information (discernment of dis/misinformation), algorithmic or cognitive bias, value-tensions and use of artificial intelligence. Co-evolution between human and social values and information technologies and education and information literacy form the framework for the model. The model can be used for design of information ethics courses at different levels of education, especially secondary level and at university level. More practically oriented courses can focus on determination of dis/misinformation, fake news or hoaxes in digital environment (social media). The role of new information professionals and academic libraries can be here significant in the framework of information literacy or copyright literacy courses. The model helps us understand and visualize decisive ethical factors of information behaviour and information literacy and can be used for further studies applied to most challenging ethical issues and dilemmas with regard to digital information.

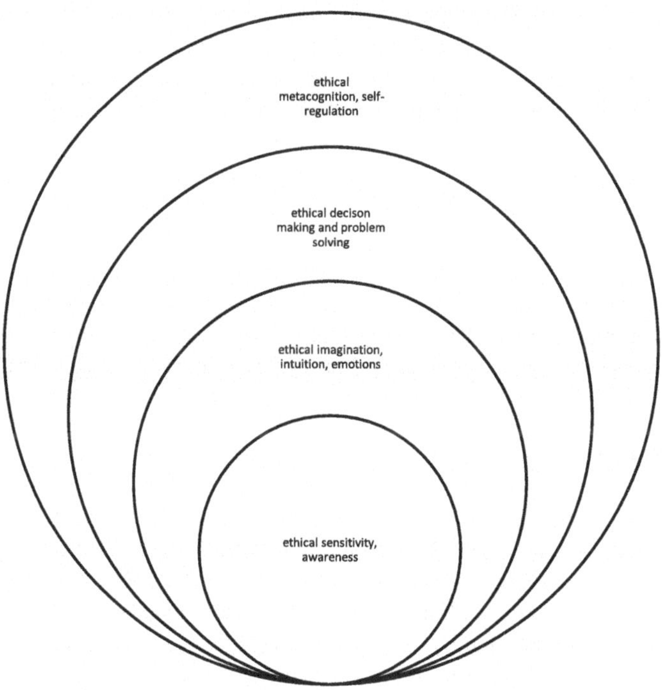

Fig. 1. A conceptual model of ethical factors of human information interactions

5 Value-Sensitive Services of Academic Libraries

Value-sensitive design as a methodology of information systems design has been developed in the framework of social informatics approximately for 20 years (Friedman and Hendry 2019). The complex approach explores moral imagination of human information behaviour and proposed several complex methods, such as value analyses, value scenarios, informed consent models, etc. The main components of the value-sensitive design are stakeholders, values and technologies and tools.

Based on our study and the value-sensitive design we have proposed a concept of value-sensitive services of academic libraries. This is an extension of our interactive model of an academic library (Steinerová 2018). We propose to design services of academic libraries in the context of value-sensitive design, based on value-analyses and value scenarios. This can result in digital portals related to ethical aspects of academic and research information use (academic integrity, ethical issues of digital publishing, research integrity, ethics of open access, data ethics and data literacy, privacy, information security and intellectual property—copyright literacy, especially antiplagiarism). A special attention should be devoted to dis/misinformation, fake news and hoaxes as part of information (academic) ethics courses. These special courses of academic libraries can be provided for students, doctoral students, researchers, teachers, administrative staff and librarians. In the academic information ecologies (Steinerová 2018) designers of value-sensitive service of academic libraries can apply value-analyses, especially identify main stakeholders (managers, students, teachers, information officers, etc.), main values of the communities, and provide them with advanced technological tools (AI tools, e.g. Chat GPT etc.) (Stahl 2021). New value-sensitive services of academic libraries should support truth and utility of information and closer collaboration of users and AI tools in information use, sharing and production. Thus, results of our Delphi study, the analyses of theoretical frameworks and the proposed conceptual model of ethical factors can be applied to design of value-sensitive services, courses and digital spaces of academic libraries. The skills related to verification of information will be a key part of information ethics courses and information literacy development.

6 Conclusions

We have analysed main topics of information ethics in published scientific literature and main topics of ethical challenges of information work by the expert consensus in a Delphi study. Results of common topics were represented by several concept models. When we compare the results, we can see that common trends and topics are represented by categories of ethics of artificial intelligence, data ethics, access and accuracy of information and ethical issues of information literacy. Differences were noted in the emphasis on selected topics. The bibliometric analyses revealed mainly topics of ethics of artificial intelligence and data ethics. The analyses of experts' discourse stressed human issues, namely ethical sensitivity and ethical awareness, education and information literacy and digital life.

We have found out that main ethical factors which shape work with digital information in information use, sharing and production are social and cultural contexts (rules, norms, traditions), social and epistemic values of information (utility and cognitive development), truth of information and verification of information in contexts of qualities of information (mis/disinformation) and value tensions between humans and advanced ICT (AI) tools in contexts of responsibility, accountability and trust. We have also found out that most significant factors of development of information ethics in future is collaboration of topically diverse different disciplines, professions and experts which can be represented by a multidimensional model of ethical challenges of digital information (cognitive, affective, social, technological, organizational, educational, community, security, legal, value-sensitive dimensions). Future development of information ethics focused on digital information will be marked by collaboration of people and intelligent tools. We have to include new concepts of "distributed morality", environmental principles of information ethics (Floridi 2013, 2014), but also risks of commercial efforts. Main driving force will be education in information ethics and information literacy of top managers and designers of information systems.

Based on our analyses and study, we have proposed a conceptual model of ethical factors of human information interactions which identified four interconnected strata of human information behaviour and information literacy, covering basic ethical sensitivity, moral emotions and moral imagination, ethical background of decision-making and problem solving, and metacognitive ethical self-management, self-regulation. The model represents the complexity and holistic principles of human information interactions, embedded in digital information environments. The model considered some visions of closer modelling of human senses and social online communication with the use of advanced intelligent technologies. We suppose that new methodological innovations can contribute to new moral frameworks for digital life, including integration of sense making with intelligent technologies. Further studies can lead to methodological innovations towards the use of AI, media, social and psychological sciences, including philosophy of information and data. Trends of development of information science are marked by divergent possible ways and we can see a kind of "fragmentary" tendencies (Vakkari et al. 2022).

This development of information ethics brought challenges for integrating research and practice of academic libraries. New models of academic libraries can include ethical factors into their value-sensitive services. That is why we have proposed ethical re-design and the concept of value-sensitive services for academic libraries based on the methodology of value-sensitive design of systems. Value-analyses can help build new digital spaces and places for special stakeholders in the academic environment. Value-sensitive services can be closer to topical ethical issues of information use and production, such as the use of rules (academic integrity, plagiarism), determination of dis/misinformation, verification of information, but also intelligent assistants in library services, and support of creative analytical information reviews, including AI tools applied to repetitive cognitive processes related to information seeking and production. Many useful intelligent services of libraries can be based on AI tools, including protection of privacy, information security, or human-robot interactions. Of course, AI tools and autonomous systems can have many risks and benefits and can be ethically challenging. But the most important tasks and challenges for visions of a new paradigm of the academic library services are still information ethics education and information literacy development (e.g. special courses on ethical sensitivity and discernment of dis/misinformation). The information ethics courses should develop the metacognitive components of information behaviour and metaliteracy which can help reflect on ethically responsible use of intellectual property, evaluate information and one's own biases, including social perception, cognitive bias, social diffusion, motivated reasoning, but also bias of algorithms. The contexts of information ethics are based on concepts of emotions, trust, responsibility, utility and truth of information. These abstract concepts can be manifested in human ethical information experience as part of use of services and products. Academic libraries should plan further ethical re-design of information systems, products, services in order to create space for more creative power of humans in information use and production.

Acknowledgement. The paper was developed in the framework of the project VEGA 1/0360/21 Social representations of ethical challenges of digital information revolution.

The author expresses her thanks to all participants of the study and collaborators who helped analyse the data.

References

ACRL.: Framework for information literacy for higher education, in *American Library Association 1996–2019*. Association of College and Research Libraries (2016). Available from: http://www.ala.org/acrl/standards/ilframework

Agarwal, N.K., Alsaeedi, F.: Creation, dissemination and mitigation: toward a disinformation behavior framework and model. Aslib J. Inf. Manage. **73**(5), 639–658 (2021). https://www.emerald.com/insight/content/doi/10.1108/AJIM-01-2021-0034/full/html

Agarwal, N.K.: Exploring Context in Information Behavior. *Seeker, Situation, Surroundings and Shared Identities*, 163 p. (Springer Nature, Cham, 2022). ISBN 978-3-031-01185-6

Bawden, D., Robinson, L.: *Introduction to Information Science* (2nd ed, 386 s). (Facet Publishing, London, 2022). ISBN 978-1-78330-495-0

Bawden, D., Robinson, L.: The dearest of our possessions: Applying Floridi's information privacy concept in models of information behavior and information literacy. J. Am. Soc. Inf. Sci. **2020**, 1–14 (2020). https://doi.org/10.1002/asi.24367

Bruce, C.: Information literacy research: dimensions of the emerging collective consciousness. Aust. Acad. Res. Libr. **47**(4), 220–238 (2016). https://doi.org/10.1080/00048623.2016.1253423

Burgess, J.T.F., Knox, E.J.M. (ed.) *Foundations of Information Ethics* (156 p). ALA, Facet, London (2019)

Capurro, R., Pingel, C.: Ethical issues of online communication research. Ethics Inf. Technol. **4**(3), 189–194 (2002). Available from: https://link.springer.com/referenceworkentry/10.1007%2F978-94-007-6046-2_127

Capurro, R.: Ethical issues of humanoid-human interaction, in Goswami, A., Vadakkepat, P. (eds.). *Humanoid Robotics: A Reference*. Dordrecht: Springer, pp. 2421–2435 (2019). Available from: https://link.springer.com/referenceworkentry/10.1007%2F978-94-007-6046-2_127

Davenport, T.D.: *The AI Advantage. How to Put the Artificial Intelligence Revolution to Work* (231 s). Cambridge: The MIT Press (2018). ISBN 978-0-262-03917-8

Fidel, R.: Human information interaction: an ecological approach to information behavior. MIT Press, Cambridge, MA (2012)

Floridi, L., et al.: AI4People—an ethical framework for a good AI society: opportunities, risks, principles and recommendations. Mind. Mach. **2018**(28), 689–707 (2018). https://doi.org/10.1007/s11023-018-9482-5

Floridi, L.: *The Ethics of Information.* (Oxford University Press, Oxford, 2013). ISBN 9780199641321

Floridi, L.: *The Fourth Revolution: How the Infosphere is Reshaping Human Reality.* (Oxford University Press, Oxford, 2014). ISBN 978-0199606726

Floridi, L.: *The Logics of Information: A Theory of Philosophy as Conceptual Design.* (Oxford University Press, Oxford, 2019). ISBN 978-0-19-883363-5

Frandsen, T.F., Sørensen, K.M.: The perceived value of the academic library: a systematic review. Inf. Res. **25**(4), 874 (2020). Accessed from: https://doi.org/10.47989/irpaper874

Friedman, B., Hendry, D.G.: *Value Sensitive Design. Shaping Technology with Moral Imagination* (229 p). (MIT Press, Cambridge (Ma), 2019). ISBN 9780262039536

Gorman, J.: Our Enduring Values: Librarianship in the 21st Century. American Library Association, Chicago (2000)

Karlova, N.A., Fisher, K.E.: A social diffusion model of misinformation and disinformation for understanding human information behavior. Inf. Res. **18**(1), 573 (2014). Available from: http://InformationR.net/ir/18-1/paper573.html

Kelly, M., Bielby, J., (eds.): *Information Cultures in the Digital Age. A Festschrift in Honor of Rafael Capurro* (479 s). (Springer, Wiesbaden, 2016). ISBN 978-3-658-14679-5. http://www.capurro.de/onres.htm.-6046-2_127

Lewandowsky, S., Ecker, U.K.H., Cook, J.: Beyond misinformation: understanding and coping with the post-truth era. J. Appl. Res. Mem. Cognit. **6**(4), 353–369 (2017). Available from: https://doi.org/10.1016/j.jarmac.2017.07.008

Lloyd, D.A.: *The Qualitative Landscape of Information Literacy Research.* Perspectives, Methods and Techniques (152 p). Facet, London (2021)

MacKey, T.P., Jacobson, T.R.: *Metaliterate Learning for the Post-truth World* (230 p). (ALA, Chicago, 2019). ISBN 978-0-8389-1776-3

Marchionini, G.: Human-information interaction research and development. Libr. Inf. Sci. Res. **30**(3), 165–174 (2008)

Nardi, B.A., O'Day, V.: *Information Ecologies: Using Technology with Heart.* (MIT Press, Cambridge, 1999). ISBN 978-02-626-4042-8

Oakleaf, M.: *The Value of Academic Libraries: A Comprehensive Research Review and Report* (182 s). (ALA–ACRL, Chicago, 2010). ISBN 978-0-8389-8568-7. Accessed from: http://www.acrl.ala.org/value

Ruokolainen, H., Widén, G.: Conceptualisation of misinformation in the context of asylum seekers. Inf. Process. Manage. **57**(3), 102–127 (2020). Available from: https://doi.org/10.1016/j.ipm.2019.102127

Saracevic, T., Kantor, P.B.: Studying the value of library and information services. Part 1: establishing a theoretical framework. J. Am. Soc. Inf. Sci. (1986–1998) **48**(6), 527–542 (1997)

Secker, J., Coonan, E.: *Rethinking Information Literacy: A Practical Framework for Supporting Learning.* Facet, London (2013)

Stahl, B.C.: *Artificial Intelligence for a Better Future. An Ecosystem Perspective on the Ethics of AI and Emerging Digital Technologies.* (Springer, London, 2021)

Steinerová, J.: *Informačné prostredie a vedecká komunikácia. Informačné ekológie* (*Information Environment and Scientific Communication. Information Ecologies*). UK, Bratislava (2018)

Steinerová, J.: Etika digitálnych informácií vo svetle sociálnych hodnôt informácií. ITlib [online]. Číslo **2**, 6–21 (2020). ISSN 1335-793X. Available from: https://itlib.cvtisr.sk/%c4%8cl%c3%a1nky/clanek3753/

Steinerová, J.: Etika tvorby informačných produktov v digitálnej revolúcii. ProInflow: časopis pro informační vědy **13**(1), 64–86 (2021). Available from: https://doi.org/10.5817/ProIn2021-1-4

Steinerová, J.: Etické faktory informačných interakcií človeka. Knihovna: Knihovnická Revue **33**(1), 5–29 (2022a). ISSN 1801-3252

Steinerová, J.: Delfská štúdia o informačnej etike: výsledky analýz diskurzu. ITLib, č. 3/4, s. 5–18 (2022b). Dostupné aj na: https://itlib.cvtisr.sk/%c4%8cl%c3%a1nky/delfska-studia-o-informacnej-etike-vysledky-analyz-diskurzu/

Steinerová, J.: Ecological and Ethical Contexts of Digital Literacy in the Light of Phenomenographic Studies. In: Kurbanoglu, S., Špiranec, S., Ünal, Y., Boustany, J., Kos, D. (eds.) *Information Literacy in a Post-Truth Era. 7th European Conference on Information Literacy, ECIL 2021 Review*. Selected Papers (pp. 157–166). (Springer, Cham, 2022). ISBN 978-3-030-99884-4. https://doi.org/10.1007/978-3-030-99885-1

Tuana, N.: Conceptualizing moral literacy. J. Educ. Adm. **45**(4), 364–378 (2007). Available from: http://site.ebrary.com/lib/uniba/reader.action?docID=10196373

Urquhart, C.: Principles and practice in impact assessment for academic libraries. Inf. Learn. Sci. [online] **119**(1/2), 121–134 (2018). Dostupné na: https://doi.org/10.1108/ILS-06-2017-0053

Vakkari, P., Yu-Wei, C., Järvelin, K.: Disciplinary contributions to research topics and methodology in library and information science—leading to fragmentations? J. Am. Soc. Inf. Sci. **2022**, 1–17 (2022). https://doi.org/10.1002/asi.24690

Walton, G., Pointon, M., Barker, J., Turner, M., Wilkinson, A.: Information discernment and the psychophysiological effects of misinformation. Glob. Knowl., Mem. Commun. **71**(8/9), 873–898 (2021). https://doi.org/10.1108/GKMC-03-2021-0052

Studying Discussion Groups and Image Boards—Ethical and Methodological Questions

Ari Haasio(✉) and Elisa Kannasto

Seinäjoki University of Applied Science, Seinäjoki, Finland
{ari.haasio,elisa.kannasto}@seamk.fi

Abstract. In this chapter, we discuss the primary methodological and ethical aspects related to studying conversations that take place on different social media platforms, such as Facebook, discussion boards, and dark web forums. The presented views are based on the authors' previous empirical research, which has focused, among other things, on social media discussion and reactions to politicians' posts, discussions on drug websites on the dark web, and the information practices of socially withdrawn (hikikomori) in picture board discussions. In addition to theory and methodological considerations, we briefly present examples based on our research. In summary, protecting anonymity and confidentiality are the most important things that should be remembered in the research on discussion groups, picture boards and, above all, the dark net.

1 Introduction

With the breakthrough of social media in the early 2000s, information behavior related to social media has been a popular research topic. Discussion groups have also been studied in information studies since the mid-1990s. The extent of the research is well illustrated by the fact that when a search is done on Google Scholar with the search terms "social media" AND "information behavior," 69 results are obtained. Search allintitle: "social media" "information seeking" on the other hand gives 210 results. In both cases, the search was targeted to the article's title. In practice, however, the number of articles related to the topic is more significant. Also, the research interest in the dark web, which with its imageboards and chats is also a part of social media, has grown (e.g., Kwon et al., 2017; Haasio & Naka, 2019; Harviainen et al., 2020; Sirola et al., 2022; Hakalahti & Harviainen, 2021).

Regarding methodology and research ethics, chats, discussion boards, and social media form separate entities with certain unique characteristics. In the following, we will discuss the issues that a researcher should consider when conducting research in these online services.

This article is based on the previous literature about this subject and the articles written by the authors. During our own research processes, we have had to consider not just the methodology but also several ethical questions due to the sensitive nature of the research. For example, when investigating drug trafficking on the dark web or issues related to the mental health and information behavior of socially withdrawn young people, careful consideration is needed. The examples in the article are based on the authors' previous research unless otherwise stated.

A. Katsirikou (Ed.): QQML 2023, 68, pp. 125–135, 2026.
https://doi.org/10.1007/978-3-032-01264-7_11

2 Content Online

The online world has ultimately changed how people communicate and continues to do so as new platforms, features, and styles develop. From online bulletin boards and text-based chats where visuals could only be built with particular keyboard characters (Suominen et al., 2019), there are now short video applications where people can communicate directly and almost in real-time by combining their own content with someone's video and then have it become a viral phenomenon. X (formerly known as Twitter) allows direct access to celebrities, politicians, and others worldwide. Thus, it is only understandable that the phenomena and discussions online have attracted researchers, developed new research methodologies, and proved a significant source of understanding the world and individuals.

Social media and all other online content include multimodal content (Salonen et al., 2021), which challenges researchers in defining research settings and methodology. With multimodal content, we mean the different combinations of text, visual, audiovisual, tags, emojis, links, and audio. Rarely would the content selected in the data include only, for example, a visual or a text. Often, these are combined, and some trends online, for example, memes, are built to include both in the message delivery. This challenges both data collection and analysis, if one analysis method is chosen and some of the content is excluded, something is already lost in the possible research outcome. This multimodal nature is not only specific to post, but in many platforms, also the comments have similar features to original posts.

The visual focus has increased with the development and increase in social media platforms. Introducing emojis as visual vocabulary (Stark & Crawford, 2015) emphasized the possibility of text becoming secondary. Memes also often play a significant role in data analysis on picture boards. They may even change the content to satirical or show their feelings differently. Lately, audiovisual content, such as TikTok videos and Instagram stories, has become increasingly popular and thus attractive also from the research viewpoint.

Engagement, the interest shown through reactions, comments, and shares of users to each other's posts, has also been a topic of many researchers because it helps to understand the algorithms and how they work—something that the platforms as companies refuse to open in detail. This has also increased research interest in emotions and how they work in online discussions (Knuutila & Laaksonen, 2019; Kannasto et al., 2023). In addition, interest in information bubbles and polarization has been increasing significantly with the increase of political communication across platforms.

3 Methodology

At its core, the focus online is on user-generated content and interaction between users because that is what defines content creation online. Both methodological decisions and ethical considerations require the researcher to understand the context of online discussions to analyze their content, nature, and significance (Turtiainen & Östman, 2013, p. 55).

3.1 Digital Research Methods

Using digital methods for data collection or analysis (see Rogers, 2024) is growing in research because social media data is rich, and manual analysis takes time and requires interpretation. Network analysis has been used quite a lot for research, for example, in discussions on X (e.g., Maireder & Schlögl, 2014). For online discussion, also sentiment analysis has become popular in detecting which topics promote negativity or positivity on social media or what kind of styles create different sentiments in engagement. For example, Knuutila and Laaksonen et al. (2020) have written about negativity promoting more negative engagement. One prominent challenge with digital methods is the difference in how prepared language models are for treating languages other than English. In addition, online discussion is highly intertextual and is often sidetracked from the main topics (Kannasto, 2021; Suominen et al., 2019). Online discussion includes flaming and trolls, and it is multimodal, so often, digital analysis is supported by at least a sample of manual coding to ensure the validity of the results.

The analysis can also be done manually. Especially when using qualitative content analysis, our experience is that an efficient method is when two researchers read the data and compare the results. Then, inter-coder reliability can be measured to ensure validity of the research and check for interpretation differences. Also, when classifying the data and using grounded theory as a theoretical framework, the reliability increases when the process is done by two researchers (see, e.g., Haasio, Harvianen & Savolainen, 2020). As a theoretical framework, grounded theory has proven to be very useful, for example, in studying the dark web, because there is not much previous research on the subject. Of course, its use depends on the problem statement, but as one option, it is a viable method.

4 Ethics

The Association of Internet Researchers published its IRE 3.0 guidelines in 2019 (Franzke et al., 2020). However, as platforms develop and new ones form and even the platforms set their own guidelines and regulations for researchers, there is a constant need for re-evaluation and updates when these guidelines are interpreted in each research setting (Markham & Buchanan, 2017, 201–202). When evaluating ethics on studying discussion groups, there are two primary starting points, (1) doing no harm, and (2) informed consent. First, the consideration of doing no harm (Mancosu & Vegetti, 2020) means that when analyzing posts on social media, the researchers should make sure that there is no information shared that is sensitive from the individual's point of view. The researchers need to consider how intimate the data is and whether the writers originally meant it to be public (Paasonen, 2013, p. 51). It has been argued that not all those who post on social media realize they are writing on a public forum (Turtiainen & Östman, 2013, pp. 49–67).

When using material published on social media, one has to be aware of its publicity because it affects the permissions needed. Pictures and writings published on social media are either entirely or limitedly public. In the first case, the forum is open to everyone and does not require a login. In the latter case, it is a closed service that requires registration or acceptance as a member. This affects how freely we can use the material. In the latter case, the use of the material is always dependent on the consent of the service administrator. In the first case, however, it can be used freely without separate permission with a careful consideration to those participating in the discussion.

When analyzing the material, the anonymity of the service must also be considered: whether it is entirely anonymous, a service that uses an avatar, or a site like Facebook where people appear with their name and picture. Even if the service is completely anonymous in principle, the chatters may tell very detailed things about themselves, and with their help, it is possible to identify the person with a fairly high degree of certainty.

In addition, the boundaries between representations of private persons and public figures, such as politicians, blur nowadays, especially on social media. Thus, it may be debatable which public pages or profiles can be evaluated as public and thus under the ethical guidelines for analysis.

However, there are also approaches in presentation that can help keep the research so that it follows the ethical guidelines. First of all, consider the need and limits of citations. In qualitative analysis, citations support the analysis and arguments but can also help trace the original posts and writers. Therefore, it needs to be considered how much of the citation is needed and sufficient. This also has to do with the data minimization principle; the amount of data gathered and analyzed must be limited, considering the need. In addition, the research framework should treat the texts as part of a whole dataset instead of an individual text tied to its author (Franzke et al., 2020). Second, anonymizing the data by removing identification information and references to individuals and only describing what is relevant to the research. This also concerns the European Union General Data Protection Regulation (GDPR) (2018), which has varying interpretations across academia.

Individual universities have their own data plans and research directives, which ensure that the research follows the GDPR and the ethical guidelines that the university has set up. For example, the approaches for researching specific social media platforms or accounts vary between universities.

Informed consent means that the individuals subjected to research should know about it and give their permission. At its most straightforward, online, this could mean that a researcher posts a question and states that those who answer the question allow their answer to be used for research and describe the research. However, this is different from how discussion groups are typically analyzed. It is more common for the researcher to define groups, discussion boards, or public accounts from which they collect comments and then analyze them after the discussions have been running for a while. In these cases, the participants have rarely been informed that the discussion will be analyzed for research purposes. However, often so many people are participating that it is impossible to inform them all. In addition sometimes the researcher may even have participated. Suppose the researcher has been involved in the discussion groups they are analyzing. In that case, it needs to be evaluated whether they have provoked or altered the direction of the discussion, thus affecting their own research data.

The next possible thing is to consider that no harm is caused to the discussion participants, and no intimate information or identification is revealed. There are challenges for this, though. Long citations may be traced with a Google search. In addition, for example memes and other visuals often carry some signature that shows whose original work they are, at least hidden in the meta information. On the other hand, finding the original work may be methodologically challenging, which may be problematic for the analysis.

As mentioned earlier, platforms have published their own guidelines regarding the use of their data for research purposes. Some allow API access, for example, through an application process, which is when automated data collection from, for example, Facebook or X may be possible. Often, these have limitations, such as no collection from individual accounts or private groups. Meta has even offered a service called Crowdtangle for researchers which has been accessible through an application process. Social media data often consists of large datasets and requires additional data storage planning, especially when it is shared in a research group.

5 Cases

5.1 The Unique Features of the Dark Web as a Source

The dark web's content differs from the open web's services. Some of the material is even illegal, such as a discussion related to the sale of drugs. If not literally illegal, the content may be questionable, immoral, or harmful. Because of the information's nature, it is especially important to handle the material very sensitively and respect anonymity (Harviainen et al., 2021).

Although Tor networks are basically anonymous, messages may contain information that can be used to trace the author's identity. Therefore, it is very important that all possible references that could reveal the author of the message or the person mentioned in it are removed from the messages analyzed and used as examples. Despite the anonymity of the dark web, the messages may even describe the person and/or their vehicle if the buyer has been deceived or robbed during the transaction (Harviainen et al., 2021).

Even if a person tries to act anonymously, he/she may leave traces on social media that make it easy to identify him/her. When one of the authors analyzed a Finnish drug website on the dark web, they often came across the same nickname (Haasio et al., 2020). The location of the nickname was revealed on which city's site he/she reported on matters related to the purchase and sale of drugs. The person also stated their age, and the nickname revealed their gender. When the researcher did a user search on social media services, the person in question was also on a social media service under the username he used on the Tor network. Most likely, it was the same person because the age, gender, and city matched, and some of the pictures had symbols referring to cannabis. The person also had recognizable facial images of themselves.

One must remember that discussion forums can be an essential channel to get peer information and support (Haasio, 2015). For example, when trying to detoxicate, drug addicts often ask for peer support and hints on how to reduce or quit using drugs. The subject is very delicate because, in Finland, using all drugs is prohibited. Asking for help from the authorities or health center is sometimes tricky because one may lose their job or even the custody of their children (Haasio et al., 2022). That explains why the discussion takes place on the dark web, not on the invisible web. People are afraid that their identity will be revealed and cause harm to them. From the researcher's point of view, it is crucial to be very careful in such a case that no identity details are revealed.

5.2 Studying Mental Health Problems

Health issues are sensitive topics; some people do not want to discuss them with their own names. Anonymity or using the dark web offers the opportunity to discuss them with confidence.

The data for our studies concerning the socially withdrawn (hikikomori) were collected from the Finnish picture board (e.g., Haasio, 2015; Haasio & Naka, 2019; Haasio & Salminen-Tuomaala, 2021a, 2021b; Haasio, 2021). In one comparative study, Japanese 4chan was also used. The picture boards used in the study differ from other social media services. Above all, memes and pictures often play a significant role. The choice of meme or image may be an essential part of the message, in some cases, the message has no text in addition to the image or meme. The use of language can also be low-style. On many other forums, for example, in Finnish Ylilauta, the language used would not be tolerated due to swearing and even some hate speech.

Exaggerating is also typical on picture boards, so researchers must consider how much the writer adds to the truth. In some cases, the discussions are flaming because there is no moderation. Flaming is a typical characteristic of especially anonymous discussion threads on social media (Suominen et al., 2019). Actually, in the Ylilauta, only Finnish legislation limits what you can talk about and how you can talk about it. In addition, anonymity lowers the threshold for bad behavior and hate speech.

Although coarse language is used on the forum of socially withdrawn young people, it is an essential source of peer support and information. Because of this, very individual health-related information, such as diagnoses and medication, is handled anonymously there. However, the researcher must note that the messages are written confidently and clearly only for their own community to read.

In the conversations held by socially withdrawn young people, the topics may also be related to other very sensitive matters, such as suicide planning or experiences of being bullied at school. The nature of the topics clearly shows that everything referring to a person must be anonymized, and the material must be handled with great sensitivity. An example of this kind of research can be mentioned in Haasio's and Salminen-Tuomaala's (2021a, 2021b) articles related to the suicide discussions of socially withdrawn young people. The material contained references to the identity of the author of a discussion group who had committed suicide, and these were removed from the material during the analysis phase.

Leaving the nickname or avatar away from the quotes is not always enough. All the possible details, which may cause identifying the person who has written the post quoted, have to be removed. Especially when the data is collected from a small country like Finland, there is always the possibility that people may know each other or recognize someone from details like living environment descriptions, hobbies or other.

5.3 Trolls

When analyzing the content of discussion group messages, it must be remembered that some of the messages may be posts made by trolls, which cannot be included in the final analysis. In identifying trolls, one critical method is to observe the language in the messages. Does the author exaggerate things? Is the content of the message excessive or malicious?

The purpose of trolls is not only to disturb and sow discord. For example, on forums dealing with social issues, trolls try to shape people's opinions in a way favorable to their own world of ideas. These people are often hired for this purpose. In some cases, governments support trolls; cases where Russian state agencies employ trolls are known (Zannettou et al., 2019).

According to Addawood et al. (2019), trolls typically use deceptive language. When analyzing the data, the authenticity of all messages must be considered. If the message is not deceptive, harsh, or exaggerated, it can sometimes be difficult to determine its authenticity. One way to get help in this process is to use artificial intelligence. There are behavior-based machine-learning approaches to identifying trolls (Alhazbi, 2020).

5.4 Politicians on Social Media

In their analysis of politician–constituent interaction during the elections (Kannasto, 2021), one of the authors has made similar notes on the discussion with Nelimarkka et al. (2015) on Twitter and Facebook. Politicians and their posts can be researched freely once they post through a politician profile or, in Twitter's case, with their profile in their political role. The same can be applied to Instagram and TikTok. Politicians are public people who also hold considerable power in society, and thus, subjecting their public representations and communication to research is allowed, whereas it is more problematic to try studying private persons' profiles on social media, which is partly happening when their comments are analyzed.

The people who participate in online discussions on politicians' or other political actors' profiles are generally private individuals, making studying these discussions more problematic regarding ethical considerations. There may be hundreds or thousands or even more individuals, so getting informed consent from all participating in the discussion can be considered impossible. The next consideration is whether their anonymity can be secured and no harm will be caused to the individuals. They are writing on a public platform. However, many may not realize it. Therefore, their communication must always be treated as a whole in research; citations should be limited, and only relevant citations should be used to support the analysis. On X (formerly known as Twitter), the data is mainly in text form, and a text-based analysis also works on Facebook discussions. However, it must be remembered that both platforms include multimodal communication forms, as stated by Salonen and colleagues (2021). This needs to be considered when planning the methodology.

When analyzing posts or discussions on more visually focused platforms, such as Instagram and TikTok, the multimodal aspect is even more crucial for the methodology. Visual analysis is essential in these platforms as part of the research design because visuals play a prominent role in communication. In addition, in TikTok, audio also plays a major role in content and engagement.

There is also a growing body of research interested in the engagement that is seen on social media from the user perspective. Here, the platform tools, for example, the reactions that are set ready and the features to share, react, or comment, are central to what the platform allows the user to do. Research on engagement often studies online viral phenomena, such as how different types of content get reactions or, for example, comments to a post. The challenge here is that the research cannot discuss the engaging ones for ethical reasons. So often, the magnitude of, for example, the network of those participating in the discussion, the individuals' background, and motives to participate are left in the dark, and it is pretty impossible to map out how many participants there are or if it is just a few who are posting multiple comments. This can be done through the netnographic research process (see Kozinets, 2015, 2020), but this requires manual analysis, which limits the amount of data that is possible to analyze. Netnographic research process could also be chosen for analyzing interaction online.

One of the authors has analyzed how the use of emojis can form communal ties between the political actor, like a politician or the party, and the constituent (Kannasto et al., 2023). Network analysis can be used to see the networks between social media users and help in understanding how these communities connect with each other. For example, information bubbles and polarization have been examined through this methodology.

5.5 Other Examples

The previous examples in our article, with the exception of political discussion and analysis, have represented very sensitive topics due to the researchers' own topic choices. The basic principles of anonymity and confidentiality are, of course, the same when researching all topics, but in some, they are emphasized even more.

A terrorist attack and stabbing of nine innocent bystanders in Turku, Finland, was even internationally noted in the media. The related research of the authors (Haasio et al. 2018) of the article and two other researchers showed well how a considerable amount of incorrect and unconfirmed information circulated on social media and, in this case, in the Facebook discussion group.

The moderated news website of the Finnish National Broadcasting Company (YLE) on Facebook was used as the source material. Commenting took place under one's own name, and several messages also referred to other people. Even though it was a moderated group, some posts contained clear hate speech. Especially when quoting the posts, anonymity was necessary, and all the hints about the personality of other people were carefully faded.

6 Conclusions

Based on our experience, the research we have done, and previous literature, trust, and anonymity are essential questions to be aware of when researching social media. Regardless of the growing body of research around the online world and social media for quite some time with a variety of topics and viewpoints, several questions keep puzzling researchers in the field. How should we consider the possibility of trolls, flaming, and disturbing and intentional provocation in discussions? Sometimes, it can be difficult to draw the line between a troll and a person who may be aggressive and over-the-top in his/her messages. What are the methods by which we, as researchers, guarantee confidentiality and anonymity? Is it ethical to study online discussions and where to draw the lines of private and public?

For the validity of the research, it is essential to weigh how much data needs to be analyzed and in what form it can be presented to ensure anonymity and do no harm to those who participate in both private and public forums. The type of forum investigated affects the questions the researcher has to remember. Are there permissions needed or not? Is the forum anonymous, is it based on the use of avatars, or do people write with their own identity? Are people thinking they are writing to a public forum, for example, when writing in a closed Facebook group, regardless of the number of people in the group, or are they participating in a discussion on X where comments are widely public? These boundaries are only sometimes clear to all social media users; therefore, researchers must take careful consideration when using this kind of data.

As new platforms arise and current ones are trying to impose restrictions on data collection and API access, researchers need to stay alert and informed. The online world develops fast, the users adopt new ways to communicate which keeps developing societal interaction and forms of entertainment. Research can help to understand these transformations and the significance of the online culture on society so it is essential to find possibilities to keep studying online discussion and develop methodologies for it.

References

Addawood, A., Badawy, A., Lerman, K., Ferrara, E.: Linguistic cues to deception: identifying political trolls on social media. In: Proceedings of the International AAAI Conference on Web and Social Media, vol. 13, pp. 15–25 (2019)

Alhazbi, S.: Behavior-based machine learning approaches to identify state-sponsored trolls on twitter. IEEE Access **8**, 195132–195141 (2020)

Franzke, A.S., Bechmann, A., Zimmer, M., Ess, C., the Association of Internet Researchers: Internet Research: Ethical Guidelines 3.0 (2020). https://aoir.org/reports/ethics3.pdf

General Data Protection Regulation: Intersoft Consulting (2018). Retrieved 15 Aug 2019. https://gdpr-info.eu Accessed 3 Jun 2020

Haasio, A.: Information seeking behaviour of the socially withdrawn. In: Madge, O.-L. (ed.) New Trends and Challenges in Information Science and Information Seeking Behaviour, pp. 23–32. Springer (2021)

Haasio, A.: Toiseus, tiedontarpeet ja tiedon jakaminen tietoverkon "pienessä maailmassa". Tutkimus sosiaalisesti vetäytyneiden henkilöiden informaatiokäyttäytymisestä. Dissertation. Tampere University (2015). https://urn.fi/URN:ISBN:978-951-44-9878-7

Haasio, A., Harviainen, J.T., Savolainen, R.: Information needs of drug users on a local dark Web marketplace. Inf. Process. Manag. **57**(2), 102080 (2020)

Haasio, A., Harviainen, T., Ylinen, A., Oksanen, A.: Tor-verkossa tapahtuvien keskustelujen merkitys huumevieroituksessa (2022)

Haasio, A., Mattila, M., Ojaranta, A., Kannasto, E.: Terrori-isku tiedontarpeiden virittäjänä: Turun puukotusten aiheuttamat tiedontarpeet. Informaatiotutkimus **37**(2) (2018)

Haasio, A., Naka, H.: Information needs of the Finnish and Japanese hikikomori: a comparative study. Qualitative Quantitative Methods Libr. **8**(4), 509–523 (2019)

Haasio, A., Salminen-Tuomaala, M.: Keskusteluryhmässä jaettavan vertaistiedon ja vertaistuen merkitys sosiaalisesti vetäytyneiden henkilöiden itsemurhakeskusteluissa. Informaatiotutkimus **40**(2), 5–27 (2021)

Haasio, A., Salminen-Tuomaala, M.: Suicide motives and protective factors-contributions from a hikikomori discussion board. Issues Mental Health Nurs. **42**(5), 417–429 (2021)

Hakalahti, M., Harviainen, J.T.: Sharing identity information on Dark Web drug boards. New Trends and Challenges in Information Science and Information Seeking Behaviour, pp. 49–58 (2021)

Harviainen, J.T., Haasio, A., Ruokolainen, T., Hassan, L., Siuda, P., Hamari, J.: Information protection in dark web drug markets research. In: Maui, H. (ed.) Proceedings of the 54th Hawaii International Conference on System Sciences, HICSS 2021, Grand Hyatt Kauai, Hawaii, USA, 4–8 Jan 2021, Tung X. Bui, Honolulu, HI, pp. 4673–4680 (2021). https://aisel.aisnet.org/hicss-54/in/social_shopping/3/

Harviainen, J.T., Haasio, A., Hämäläinen, L.: Drug traders on a local dark web marketplace. In: Proceedings of the 23rd International Conference on Academic Mindtrek, pp. 20–26 (2020)

Kannasto, E., Laaksonen, S.M., Knuutila, A.: I 🌷 🍀FI You!-Emojis as emotional-political signifiers in Finnish election campaign discussion online. In: Proceedings of the 56th Hawaii International Conference on System Sciences (2023). https://hdl.handle.net/10125/102925

Kannasto, E.: I am horrified by all kinds of persona worship!: Constructing Personal Brands of Politicians on Facebook. Dissertation, University of Vaasa (2021). https://osuva.uwasa.fi/bitstream/handle/10024/13161/978-952-476-983-9.pdf?sequence=2

Knuutila, A., Laaksonen, S.-M.: Viraali vihaisuus ja tahmea nauru: tunteet ja algoritmit digitaalisessa vaalikampanjoinnissa. In: Borg, S., Kestilä-Kekkonen, E., Wass, H. (toim) Politiikan ilmastonmuutos: Eduskuntavaalitutkimus 2019. Oikeusministeriön julkaisuja. Selvityksiä ja ohjeita, Nro 2020:5, Oikeusministeriö, Helsinki, Sivut 394–418 (2020). http://urn.fi/URN:ISBN:978-952-259-838-7

Kwon, K.H., Priniski, J.H., Sarkar, S., Shakarian, J., Shakarian, P.: Crisis and collective problem solving in dark web: an exploration of a black hat forum. In: Proceedings of the 8th International Conference on Social Media & Society, pp. 1–5 (2017)

Kozinets, R.V.: Netnography: Redefined. Sage (2015)

Kozinets, R.V.: Netnography: The Essential Guide to Qualitative Social Media Research, 3rd edn. Sage (2020)

Mancosu, M., Vegetti, F.: What you can scrape and what is right to scrape: a proposal for a tool to collect public Facebook data. Social Media + Society **6**, 3 (2020). https://doi.org/10.1177/2056305120940703

Maireder, A., Schlögl, S.: European Parliament Members´Twitter Networks (2014). Retrieved 17 Jan 2015: http://homepage.univie.ac.at/axel.maireder/2014/03/mep-twitter-networks/

Markham, A., Buchanan, E.: Research ethics in context. Decision-making in digital research. In: Schäfer, M.T., van Es, K. (eds.) The Datafied Society. Studying Culture Through Data. Amsterdam University Press, Amsterdam (2017)

Nelimarkka, M., Laaksonen, S.-M., Tuokko, M., Valkonen, T.: Platformed interactions: how social media platforms relate to candidate-constituent interaction during Finnish 2015 election campaigning. Social Media + Society **6**(2), 1–17 (2020). https://doi.org/10.1177/2056305120903856

Paasonen, S.: Ihmisiä, kuvia, tekstejä ja teknologioita. In: Laaksonen, S.-M., Matikainen, J., Tikka, M. (eds.) Otteita Verkosta, pp. 34–48. Vastapaino, Tampere (2013)

Rogers, R.: Doing Digital Methods, 2nd edn. Sage (2024)

Salonen, M., Kannasto, E., Paatelainen, L.: Sosiaalisen median kommenttien analyysi multimodaalisesta näkökulmasta digitaalisen journalismin ja poliittisen viestinnän tutkimuksessa. Informaatiotutkimus **40**(3) (2021)

Sirola, A., Nuckols, J., Nyrhinen, J., Wilska, T.A.: The use of the Dark Web as a COVID-19 information source: a three-country study. Technol. Soc. **70**, 102012 (2022)

Stark, L., Crawford, K.: The conservatism of emoji: work, affect, and communication. Social Media + Society **1**(2) (2015). https://doi.org/10.1177/2056305115604853

Suominen, J., Saarikoski, P., Vaahensalo, E.: Digitaalisia kohtaamisia: Verkkokeskustelut BBS-purkeista sosiaaliseen mediaan. Gaudeamus, Helsinki (2019)

Turtiainen, R., Östman, S.: Verkkotutkimuksen eettiset haasteet: Armi ja anoreksia. In: Laaksonen, S.-M., Matikainen, J., Tikka, M. (eds.) Otteita Verkosta, pp. 49–67. Vastapaino, Tampere (2013)
Zannettou, S., Caulfield, T., Setzer, W., Sirivianos, M., Stringhini, G., & Blackburn, J.:. Who let the trolls out? Towards understanding state-sponsored trolls. In: Proceedings of the 10th ACM Conference on Web Science, pp. 353–362 (2019)

The Value of Alliances Among Libraries and Their Cultural Management: The Case of the Maritime Libraries Network

Stavroula Kouri[1](✉) and Anthi Katsirikou[2]

[1] National Technical University of Athens Library & Information Center, Zografou, Greece
s_kouri@mail.ntua.gr

[2] University of Piraeus Library, Piraeus, Greece

Abstract. The constantly changing economic and social environment, in connection with the recent epidemiological crisis, imposes the focus of libraries and cultural organizations on their sustainability. In this context, alliances are an important resource along with an integrated cultural management strategy, adapted to the ever-changing environment. The case study focuses on collaboration between libraries of heterogeneous bodies with a common theme, which is a modern and innovative form of consortium, as the common object addressed by the participating heterogeneous members, contributes to the development and planning of joint actions. The object of this presentation is the creation of strategic planning of Maritime Libraries Network (MarLiNet), a collaborative scheme between university libraries and private libraries of cultural institutions. In the context of this process, the analysis of the internal and external environment of MarLiNet was carried out, using the PESTLE and SWOT analysis to determine the objectives of the Network for the drawing up of the strategic planning. The research process was based on the positions and statements of the members of MarLiNet and in the observance of specific international methods of strategic analysis, to render with the greatest possible clarity and accuracy the strategic planning of the case study. The objective purpose of the strategic planning is to capture as effectively as possible the long-term direction of the cooperative network and to examine and study the future progression of MarLiNet.

Keywords: Cultural Institution · Libraries · Cultural Management · Alliance

1 Introduction

In the modern digital era, sharing knowledge and acquiring new skills is closely tied to collaboration and communication. As key institutions that have traditionally supported the dissemination and organization of knowledge, libraries need to adapt to the prevailing trends of cooperation and interconnectedness across various themes, regions, or globally.

Collaboration among public, academic, or municipal libraries, along with networks involving both public and private libraries, is increasingly prevalent today. Similarly, consortia formed around shared interests or subjects represent a newer type of partnership,

A. Katsirikou (Ed.): QQML 2023, 68, pp. 136–155, 2026.
https://doi.org/10.1007/978-3-032-01264-7_12

where the common focus among participants drives the creation and implementation of collective strategies.

The Maritime Library Network is a novel initiative aimed at fostering collaboration among libraries, universities, and private organizations focused on maritime-related fields such as shipping, marine science, naval engineering, and ship construction. Originating as an idea in 2010, it officially launched in 2018 with the involvement of eight founding member libraries. Although the network lacks a formal strategic plan, it has outlined a vision and goals, yet its purpose remains somewhat vague, and specific objectives have not been clearly defined.

The main thrust of this article is based on a thesis prepared in the framework of a postgraduate program of the Hellenic Open University entitled "Management of Cultural Units" with supervisor Dr. Anthi Katsirikou. At its core, the research focused on examining both the external and internal contexts of the case study, followed by the formulation of a strategic plan for the Maritime Library Network. This was achieved through insights gathered from analyzing these two aspects.

1.1 Main Goal and Objectives

The primary objective of this research is to underscore the significance of synergistic relationships between libraries and other cultural entities, using the Maritime Libraries Network (MarLiNet) as a prime example. To achieve this, several key objectives have been identified such as the investigation of the efficacy of synergies within organizations that possess diverse characteristics, the incorporation and application of the best practices regarding synergies observed in Greece and internationally within the context of our case study and finally the formulation of a strategic plan specifically tailored for the Maritime Libraries Network (MarLiNet).

Developing a comprehensive strategic plan for this thematic network presents a significant challenge for enhancing cooperative efforts among its members and ensuring its longevity.

The overarching goal is to integrate and apply international best practices for managing synergies within actual collaborative networks involving cultural organizations and libraries. Additionally, it aims to identify and incorporate contemporary trends in cultural management and international practices into the case study, specifically focusing on the Maritime Libraries Network (MarLiNet). Another objective is to select the most suitable management and administrative approach for MarLiNet Furthermore, it offers a unique opportunity to introduce strategic planning methodologies into other library collaborations in Greece.

1.2 Methodology

This research employs a comprehensive review of both global and local academic materials, utilizing a wide range of resources including scholarly articles, periodicals, textbooks, and online research findings. The core research technique involves conducting a detailed case study, which is primarily executed through qualitative analysis. Data for this case study is sourced from both primary and secondary information channels, complemented by a variety of methodologies like interviews, direct interactions within the collaborative network, observations, and examination of the network's operations and outputs. The research strategy adopted is action research, aiming to bridge theoretical knowledge with practical application in shaping the study framework. Qualitative methods are specifically utilized to delve into the nuances of the study paradigm, capturing insights into its conceptual underpinnings, viewpoints, or experiential aspects. This approach facilitates a deeper understanding of the network's dynamics and fosters innovation in research directions. Additionally, the research incorporates secondary data derived from various textual, visual, or video-based representations of the case study.

1.3 Challenges

There were several challenges encountered during the implementation of the action research methodology, such as the difficulty of creating a strategic blueprint for a library collaboration, especially considering the scarcity of established strategic plans for Greek library collaborations despite numerous efforts to formulate such plans for individual libraries. The research process was significantly hindered by preventive measures taken in response to the pandemic, including the imposition of a general curfew and the temporary closure of libraries belonging to MarLiNet members, which restricted physical interaction among network members. A further challenge encountered was the diversity among network members, given that alliances between varied organizations and institutions often struggle with reconciling differing values, principles, corporate cultures, and user engagement strategies.

2 Basic Concepts and Principles of Synergy Management

2.1 The Concepts of Cultural Organization and Library

In the context of understanding the value of synergies between heterogeneous cultural organizations and libraries, it is useful to analyze some basic concepts to demonstrate the connection between them and their effective use in the theoretical and research approach of the subject.

Cultural organizations can be viewed through two lenses: a narrower perspective and a wider one. In the narrower view, these entities are defined by their role in producing and sharing artistic creations, focusing on either performing arts (like symphony orchestras, theaters, operas, etc.) or visual arts (including art galleries, museums, libraries, and cultural preservation efforts). Expanding this definition, the broader interpretation encompasses what is known as the cultural sector, which extends beyond traditional artistic outlets to encompass industries such as film production companies, movie theaters, broadcasting services (television and radio), and other forms of media [31].

The perception of cultural organizations by the public is shaped by three primary factors derived from management and organizational sciences, as outlined by Paquette and Redaelli in 2015 [30]. These factors are:

- The legal framework within which these organizations operate.
- The structure of dividing labor among different departments or institutions.

- The perspective of viewing organizations as open systems interacting with their environment.

Linking all three approaches, cultural organizations are those that, regardless of the institution to which they belong, have as their main objective the fulfilment of their basic mission, using as a basic weapon collectivity and continuous feedback with the important aim of attracting and shaping citizens with a well-developed cultural aesthetic.

Libraries on the other hand, serve as both repositories of books and other printed materials, carefully organized, and preserved for public use, and physical spaces—rooms, buildings, or facilities—that house these collections. Beyond being places for accessing literature, today's libraries have evolved into vibrant social hubs hosting events like concerts, exhibitions, and book readings, while continuing to be cherished study areas for individuals seeking knowledge and learning [4].

The library is perceived as a dynamic and evolving organism, continuously adapting to changing circumstances and swift advancements, much like a kaleidoscope that transforms with each twist, presenting fresh arrangements and patterns. This concept is ever-present and actively reshapes itself, embodying an ongoing cycle of renewal and transformation [4].

Libraries are divided into the following main categories or types:

- Academic Libraries: These libraries serve colleges, universities, and other higher education institutions. They offer a vast collection of books, journals, and digital resources, primarily aimed at students and faculty for research and academic purposes. Academic libraries often have specialized collections in various disciplines and support a wide range of research activities.
- The National Libraries, which usually function as national repositories, and where copies of all the country's publications are deposited, according to legislation. In Greece, it also comes under the Ministry of Education.
- Public Libraries: Public libraries are accessible to anyone who wishes to use their facilities and borrow materials. They provide a wide array of services, including lending books, offering computer access, providing programs for children and adults, and serving as community centers. Public libraries play a crucial role in promoting literacy, lifelong learning, and community engagement.
- Special Libraries: Special libraries cater to specific groups or interests, such as law firms, hospitals, corporations, or associations. These libraries collect and organize information tailored to the needs of their users, which may include legal documents, medical texts, business reports, or historical records. Special libraries are designed to support the mission and objectives of their parent institution.
- School Libraries: School libraries are integral parts of educational institutions, from elementary schools to high schools. They provide students with access to books, online resources, and other educational materials. School librarians facilitate learning by teaching information literacy skills, assisting with research projects, and encouraging reading for pleasure.

Lastly, there are digital libraries or virtual libraries, which are not really a separate category of libraries, as they have no legal status, since they are mainly collections of the above-mentioned types of libraries and are in fact a consequence of the technological evolution of the time.

Libraries prioritize several key values, including the excellence of their services, easy access to information, upholding privacy, fostering creativity, and enhancing the overall experience. These values are continuously assessed and appreciated by both library users and visitors.

2.2 Management of Libraries and Cultural Organizations

Cultural management diverges significantly from the administration and operation of profit-driven companies. At its essence, for-profit entities are designed to generate income, whereas non-profit organizations, particularly those within the cultural sector, aim to underscore the importance of specific social requirements and serve as representatives of public interests, either directly or indirectly [29].

Cultural organizations must grapple with critical inquiries such as defining their purpose, assessing their present actions, identifying the core components of their mission, determining their target audience, understanding public perception of their role, evaluating their actual contributions to the community and culture, and articulating their values and priorities. Essentially, the responses to these queries form the foundation of a cultural organization's mission, vision, and values, thereby facilitating discussions within the local community regarding the arts and culture and how they can enhance the overall quality of life.

The basic function, which is the main study component of this research, is the planning or programming function. As McNamara [28] suggests, planning involves determining your destination, the reasons behind wanting to reach it, the strategies to achieve it, the resources required, and the criteria to assess if the goal has been met. This makes it clear why planning is frequently seen as a navigational tool guiding the journey towards success.

To facilitate planning and programming within cultural organizations without compromising on the effectiveness of planning, Kaiser [18] outlines a strategic planning process divided into three stages:

1. Foundation/Positioning: Establishing the strategic framework that underpins the planning process and crafting a compelling mission statement that energizes the entire strategy.
2. Analysis: Gathering information and conducting both internal and external assessments to uncover the primary strategic challenges that need addressing during the design phase.
3. Strategic Implementation: Formulating artistic, managerial, and financial strategies that are most likely to fulfill the organization's mission effectively.

2.3 The Concept, the Types of Synergy, the Development Axes

Collaboration, etymologically speaking, signifies working collectively, implying that there is an activity or project undertaken by more than one individual. This term has been utilized across various domains for centuries, long before the advent of modern science. Its origins trace back to the ancient Greek word "synergy," denoting collective effort and cooperative action.

Across international and national contexts, there are many instances of synergy between organizations. Some of the most prevalent forms of collaboration, as defined, include:

- Networks: Composed of individuals connected through various types of relationships, ranging from formal to informal structures.
- Coalitions: Groups of organizations united by a shared objective and commitment to collaborative decision-making to affect external bodies or goals.
- Movements: Collective efforts aimed at achieving social transformation through a unified approach and long-term objectives.
- Strategic Alliances: Collaborative endeavors between independent organizations aiming to achieve a mutual goal without compromising their autonomy.
- Public-Private Partnerships: Collaborations between public and private entities to provide specific services or advantages.

Creating synergies among varied organizations encompasses seven crucial steps. While the following model isn't intended as a blueprint for forming a consortium or collaborative network of non-profit organizations per se, it serves as a valuable guide to the complexities involved in such collaborations. McLaughlin [27] outlines these seven pivotal actions:

- Initiation, Exploration, and Analysis: Starting the process, exploring options, and analyzing potential outcomes.
- Team Composition and Planning: Assembling the right team and outlining a strategic plan.
- Setting Common Objectives: Agreeing on shared goals to guide the collaboration.
- Establishing an Organizing Committee: Creating a leadership structure to oversee the collaboration.
- Pursuing Quick Achievements: Focusing on early wins to build momentum.
- Institutionalizing the Market: Integrating the collaboration into existing systems and practices.
- Implementation and Evaluation: Executing the plan and regularly assessing progress and outcomes.

The synergies between libraries and cultural organizations encompass a wide range of interactions, including the convergence of museums with libraries and archives, known as Memory Institutions. This collaboration facilitates resource sharing, enhances diversity and variety, and includes various types of library synergies, extending beyond just libraries to involve other organizations.

2.4 Principles and Factors of Synergy Success

Hughes and Weiss [14] outline five key principles for effective collaborative management:

1. Focus on Process Over Plan: It's crucial that collaborative initiatives concentrate more on the execution of the collaboration rather than detailing a rigid action plan. A formal plan might overshadow the necessity to delve into and define the dynamics of the working relationship among collaborators, focusing not just on what they'll do but also on how they'll interact.
2. Synergy Evaluation: Developing an assessment of synergy that considers not only the alliance's goals but also its progression and evolution.
3. Leveraging Differences: Striving to minimize differences among participants and utilizing them to add value. Organizations become allies due to their unique differences, including expertise, processes, and cultures, which they aim to capitalize upon.
4. Promote Collaborative Behavior: Encouraging behaviors that strengthen the working relationship from the outset of the alliance and sustaining this throughout the collaboration.
5. Commitment to Audience Needs and Relationship Management: Ensuring that the collaboration meets the needs of the audiences served by the participating organizations and effectively manages the relationship between them.

Drawing on insights from the business sector, where the creation of organizational alliances is extensively studied, Faulkner [5] identifies four crucial elements that contribute to the success of alliances:

- Positive Attitudes Among Members: The presence of a positive mindset among alliance partners is vital.
- Clear Organizational Arrangements: Having well-defined structures for the alliance ensures smooth operations.
- Philosophy of Organizational Learning: Embracing a culture of continuous learning within the alliance.
- Synchronized Long-Term Goals: Aligning the long-term objectives of the alliance partners.

3 Synergies and Cultural Management

Synergy management or collaborative management is characterized as the method of integrating through the establishment of an efficient collaborative implementation mechanism. A successful collaborative management model necessitates a robust programmatic plan backed by a comprehensive agreement. Key principles of a successful inter-organizational alliance include assessing the value brought by the alliance, finding common ground with partners, committing to managing the interface with them, and setting up formal systems and structures. Hughes and Weiss [14] emphasize these aspects as crucial for the success of alliances.

3.1 Examples from the International and National Area

Determining the exact historical timeline of synergies between libraries is challenging due to varying accounts in the international literature. Blackburn [3] traces the roots of library collaborations back to the Ancient Greeks, highlighting early collaborations in Syria and Mesopotamia prior to colonial invasions. Conversely, Grisham [7] marks the inception of library collaborations in the United States during the early 19th and 20th centuries, specifically noting the Library of Congress' introduction of the library card in 1901.

The saving of resources, the exchange of experience and know-how, the sharing of sources and resources are among the primary benefits of library partnerships, which make them dominant and particularly necessary. The following are indicative examples of alliances between libraries and/or cultural organizations at international level:

- International Alliance of Research Library Associations (IARLA). The IARLA Alliance serves as a collaborative space for its member organizations to voluntarily unite and shape consensus positions aligned with the evolving global agenda of research libraries and open academic knowledge. Governed by its initial five founding research library organizations, the Alliance functions as an informal network without formal membership requirements or associated costs [16].
- International Association of University Libraries (IATUL). Founded in Düsseldorf, Germany, in May 1955, the International Association of Technology University Libraries (IATUL) emerged as an international forum dedicated to the exchange of ideas concerning libraries in technology universities globally. Initially, IATUL was a voluntary, non-governmental organization comprising libraries represented by their directors responsible for information services and resource management. By 2014, at its annual conference in Helsinki, the IATUL General Assembly broadened its scope to encompass all university libraries worldwide, regardless of their academic focus. Consequently, the organization's name was revised to the International Association of University Libraries [17] to reflect this expansion.
- Literacy for All—European Library Network. Is a specialized thematic network focused primarily on serving the needs of public and community libraries, particularly in the realm of adult education. Established in 2014 with 20 founding members across Europe, the network aims to foster discussion and exchange ideas regarding functional literacy and the significant role libraries can play in enhancing the literacy levels of Europeans [19].
- Health Sciences Library Network of Kansas City (HSLNKC). The Health Sciences Library Network of Kansas City (HSLNKC) aids medical librarians and healthcare practitioners in enhancing health instruction and overall wellness across the broader Kansas City region. These librarians offer crucial information services to various entities such as hospitals, educational institutions, businesses, research facilities, and community groups. -The network encompasses 26 libraries associated with health-related institutions and organizations within Kansas City, USA [13].

In Greece, library synergies began tentatively at the end of the 1990s, with the Hellenic Academic Libraries Link (HEAL-Link) consortium being a typical example. While synergies in the international arena had begun several decades earlier, they made their presence felt in Greece with the onset of the economic crisis that hit the country at the end of 2000. In recent years, however, in Greece, there has been a great growth in thematic collaborative networks, where an indicative presentation of some of them follows.

- International Association of Music Libraries, Archives and Documentation Centres Greece (IAML-GR) [15]. Established in 2016, the Hellenic Branch of the International Union of Music Libraries, Archives, and Documentation Centres (IAML-GR) operates as an autonomous nonprofit entity. Its foundation was motivated by the desire to unite music libraries and archives, along with their respective professionals—librarians, archivists, information scientists, and musicologists—to enhance the availability and understanding of music information. This initiative aims to enrich music education within Greek society. From its inception, IAML-GR has focused on elevating awareness and utilization of Greek music collections and supporting those who engage with these resources. The organization comprises both individual members, primarily librarians and archivists specializing in music, and institutional members managing music collections and archives. These include academic library departments, the National Library, music archives, conservatories, and associations, whether public or private (IAML-GR).
- Hellenic Economic Library Network (H.E.LI.N). Founded in 2012, the Hellenic Economic Library Network (H.E.LI.N.) is a consortium comprising 10 members, including university libraries, information centers, and research and government institutions located in Attica. These members collectively own, manage, and provide access to collections, resources, and information services in the field of economics. The network represents a cooperative alliance among various types of libraries, including three public academic libraries (Athens University of Economics & Business, University of Piraeus, Harokopio University of Athens), two private specialized libraries (Alpha Bank, National Bank), two public specialized libraries (Bank of Greece, Ministry of Economy & Finance), two specialized libraries of independent institutions (Centre of Planning and Economic Research, Hellenic Statistical Authority), and one research library (Piraeus Bank Group Cultural Foundation) [12].
- Greek Libraries Network/National Library of Greece. Launched in 2017, the Network of Greek Libraries under the National Library of Greece encompasses 254 libraries across various types and classifications nationwide. Its foundational aim is to bolster Academic, Research, Public, Municipal, and School Libraries, aiming to enhance, grow, and broaden the services they offer to their patrons. Predominantly, the Network includes Municipal libraries spread across nearly every region of the country, alongside several Children's libraries from diverse locations, Private School libraries, Infant-Toddler libraries, and Youth and Teenage libraries [6].

4 Comparative Analysis of an International and a Greek Consortium of Academic Libraries

To gain insights into the significance, functioning, and efficacy of library consortia both internationally and nationally, this discussion attempts a comparative analysis between two academic library consortia. One example is the M25 Consortium of Academic Libraries, headquartered in London, UK (M25 Consortium of Academic Libraries), and the corresponding Greek consortium of Academic Libraries (HEAl-Link).

4.1 Presentation of the Consortia

The M25 Consortium of Academic Libraries is a collaborative organization that works to enhance library and information services within the M25 region and beyond, encompassing 55 member institutions including higher education, national, museum, and other scholarly libraries. Their goal is to offer services and resources beneficial to library staff, learners, and researchers. The M25 Consortium provides various benefits to its members, such as staff training and development programs, access schemes for postgraduate research students and staff, business continuity and disaster recovery resources, networking events, annual conferences, director briefings, participation in governance groups, staff development awards, task and finish groups and speed meet introductions [20].

Originally envisioned in the early 1990s, the Consortium of the Association of Hellenic Academic Libraries emerged as a collaborative initiative to address shared challenges such as securing funding, continuous staff development, navigating legal and administrative hurdles, and embracing technological advancements. Comprising 43 member libraries, this consortium facilitates access to comprehensive electronic journals, books, and bibliographic databases via its web portal for members of the Association of Greek Academic Libraries. Beyond providing essential digital resources, HEAL-Link now extends its offerings to academic libraries and their patrons with a variety of enhanced services [8]:

- AMELib: An Accessible Multimodal Electronic Library, offering a broad spectrum of accessible content.
- HEAL-Legal: A service providing legal advice on intellectual property, copyright, and personal data matters.
- HEAL-Link's Information Literacy Educational Platform: An online educational platform designed for undergraduate, postgraduate students, and teaching staff to enhance information literacy skills.
- Scientific Communication: Support for the scholarly communication requirements of HEAL-Link institution members.
- "Kallipos" Repository: Aimed at systematically documenting, organizing, and preserving academic and research texts and learning materials for long-term accessibility.

4.2 Comparative Analysis

While these two consortia share similarities, they also exhibit distinct features. This analysis seeks to delve into both the likenesses and differences between them, focusing on aspects such as the number of members, funding mechanisms, organizational structure, management approach, collaborative collection development strategies, nature of collections, resource and service sharing practices, and more. The goal is to uncover the strengths and weaknesses inherent in each consortium's design and operations.

As outlined in Table 1, here's how the comparison of similarities between the two consortia is presented:

1. Membership Composition: There isn't a significant disparity in the number of members between the two consortia; they are roughly on par. Geographically, the distribution of members across various regions within their respective countries is comparable for both consortia. Additionally, neither consortium exclusively includes academic

Table 1. Similarities among M25 and HEAL-Link

Comparison category	M25	HEAL-Link
Members (number and affiliation)	55	43
	Academic institutions + collaborative institutions	Academic institutions + collaborative institutions
Administration	1 representative with voting right to vote + Administrative Board elections	1 representative with voting right + Administrative Board elections
Income	Membership subscriptions	Membership subscriptions
Operation serving community	General Assembly Academic Community	General Assembly Academic Community

libraries among its members; they also engage with other types of institutions that aren't solely academic, such as research libraries and museums in M25, and research institutes in HEAL-Link.

2. Both consortia adhere to a uniform governance structure, allowing each member institution to nominate a representative, typically the head of the member library, who possesses full voting privileges. They feature a similar Board of Directors, chosen by the General Assembly, with the primary distinction being the frequency of elections: M25 conducts annual elections, whereas HEAL-Link's Board serves for three-year terms.
3. Financially, both consortia sustain themselves through membership dues paid by their members, with M25 benefiting additionally from sponsorships, a revenue stream does not present in HEAL-Link.
4. As previously discussed, the principal decision-making body within both consortia is the General Assembly, which operates through voting mechanisms, selects the Board of Directors, and facilitates discussions on various topics, including proposals and ideas.
5. Contrarily, the scope of service offered by the two consortia significantly diverges based on the community they cater to. Specifically, the focus is on the academic communities within the member universities, encompassing faculty, students, and administrative personnel.

Table 2 outlines the distinctions between the two consortia as follows:

1. M25 has implemented a strategic plan named "The M25 Consortium Strategy 2019–2022: Connecting Libraries, Staff Development." This plan culminates with an Annual Action Plan for the last year, 2022, which encapsulates the consortium's accomplishments during 2020–21 and outlines its future activities for 2021–22 [21]. Conversely, HEAL-Link lacks a defined strategic plan and does not provide comprehensive information regarding its vision, mission, and values on its website.
2. The goals and purposes of the two consortia vary considerably. Specifically, M25 aims to motivate and enhance staff development, foster collaboration through its network, and achieve recognition via its collections and services [21]. In contrast,

Table 2. Differences among M25 and HEAL-Link

Comparison category	M25	HEAL-Link
Administrative model	Defined strategic planning	Luck of Strategic Planning (Program Agreement)
Strategic goals	Library connection, Staff Development	Electronic Resources Portal
Services	Staff training and development program Access to teaching and research collections Simultaneous Directory Search Program	Unified electronic source search engine Open access agreements Accessible Multimodal Electronic Library (AMELib)

HEAL-Link, despite lacking a public declaration of its strategic goals and objectives, primarily focuses on aiding the scientific community by offering access to electronic resources, including full texts of journals, books, and bibliographic databases, through its web portal. Thus, the strategic objectives of M25 emphasize staff development and networking, while HEAL-Link concentrates on facilitating scientific research by providing a broad array of electronic resources.

3. The third distinction lies in the services offered by each consortium. M25 offers several key services to its members:
   (a) cpd25: A staff development and training program designed for all levels of library staff within the M25 Academic Library Consortium, aimed at enhancing professional skills and knowledge [22].
   (b) Access25: This program facilitates reciprocal access to educational and research materials across more than 50 libraries located in London and the Southeast of England, promoting shared resources among members [23].
   (c) Search25: A tool that enables users to search across the catalogs of most M25 Consortium libraries, providing real-time information on the availability of materials and access to library services. It also offers assistance for issues encountered in individual libraries [24].

HEAL-Link distinguishes itself from M25 by focusing on services that cater to different needs. Unlike M25, which emphasizes staff development and networking, HEAL-Link does not prioritize staff skill enhancement nor offer a unified platform for accessing the institutional repositories of its member libraries. Instead, HEAL-Link has developed an integrated automated system (ILSaS) involving 26 consortium member libraries, consolidating catalog data from all participants. Additionally, HEAL-Link provides a centralized search engine for its electronic resources, including a vast collection of e-journals, databases, and e-books [9]. Beyond this, HEAL-Link engages in open access agreements, aiming to facilitate the publication of papers in open access formats alongside traditional subscription-based access to scientific materials [10]. A unique service offered by HEAL-Link is AMELib, designed to make academic library content accessible to print-disabled users, enhancing inclusivity through a user-friendly web application [11].

4.3 Advantages and Disadvantages

Below are the advantages of the M25 and Heal-Link synergies:

1. Strong Governance: Both consortia operate under the guidance of educational institutions, with M25 being overseen by governmental entities responsible for higher education and HEAL-Link under the Greek Ministry of Education and Religious Affairs. This governmental oversight guarantees consistent subscription and service provision due to their backing.
2. Financial Stability Through Member Contributions: The primary source of funds for these consortia comes from member fees. This model not only grants privileges to core members but also imposes a responsibility on them to financially support their consortium, encouraging all members to address gaps in public funding.
3. Defined Organizational Structure: Each consortium employs a comparable management structure, featuring a board of directors elected by members and a General Assembly for decision-making. This setup effectively engages all members across regions and facilitates collective decision-making.
4. Diverse Functions and Continuous Growth: The M25 and HEAL-Link consortia are versatile, continuously enhancing their networks of collaboration among members and expanding their library-related services.

The main disadvantages of the M25 and HEAL-Link consortia are as follows:

1. Inadequate Service Offerings:

Both consortia fall short in offering certain crucial services, though they cater to different user needs. For instance, M25 excels with a comprehensive search feature across all member directories, showcasing its effectiveness and relevance to community needs. However, it lacks a comparable service for accessing electronic resources, a critical need during the recent pandemic when digital tools were vital for research. Conversely, HEAL-Link focuses heavily on electronic resources but overlooks staff training, member education, and integrating its catalogs for streamlined searching.

2. Narrow Focus and Objectives:

A detailed comparison highlights the challenge posed by the narrow scope of objectives within both consortia. The narrow focus often restricts the potential for addressing existing limitations and exploring new opportunities.

4.4 Conclusions

From the comparative examination conducted, it becomes evident that there are no significant disparities in the organizational structures between international academic library consortia like M25 and domestic Greek consortia such as HEAL-Link. It could be argued that the Greek consortium aligns better with technological advancements and the evolving demands of academia, given its emphasis on facilitating research through digital platforms.

However, a notable distinction lies in the strategic direction of the Greek consortium. Unlike the absence of a clearly outlined strategy in some cases, the core member libraries of HEAL-Link have developed a formal agreement outlining the consortium's goals, formation process, member rights and responsibilities, the authority of the General Assembly, and other operational aspects. This structured approach was formalized through the HEAL-Link Programmatic Agreement, which was ratified and endorsed by all university rectors involved in the consortium in May 2020.

5 Case Study—Maritime Libraries Network (MarLiNet)

In the realm of examining and comprehending the cultural administration of collaborations and alliances between libraries and a variety of cultural organizations and agencies, this section endeavors to outline a strategic plan for a thematic network of libraries in Greece. This network, known as the Maritime Libraries Network (MarLiNet), encompasses the vast majority of systematically organized maritime-focused collections found within special and university libraries across the country.

5.1 Profile of MarLiNet

MarLiNet was established in 2010, originating from a collaborative initiative involving three libraries: the Eugenides Foundation Library, the Ekaterini Laskaridis Foundation Library, and the University of the Aegean Library. These institutions shared a common interest in collections pertaining to maritime sciences. This initial concept evolved into a full-fledged thematic network of libraries across Greece, officially launching in early 2018 as the Maritime Libraries Network (MarLiNet). The collections related to shipping and marine sciences have seen significant growth, expanding in 2018 with the addition of five more members and further growing in 2021 with the inclusion of two more members, resulting in a total of ten members. This expansion encompasses a variety of library types, including five special libraries, five university libraries, four private libraries, and six public libraries, as the respective libraries of the following organizations participate [25]:

1. Aikaterini Laskaridis Foundation
2. Eugenides Foundation
3. Hellenic Institute of Marine Technology
4. Hellenic Maritime Museum
5. Hellenic Naval Academy
6. Hellenic Naval War College
7. National Technical University of Athens
8. Naval History Department (Hellenic Navy)
9. University of the Aegean
10. University of Piraeus

5.2 Objectives–Operation–Actions

The primary objective is to enhance services and maritime collections. The aims include expanding material accessibility, improving the readability of maritime collections, and maximizing the reach of MarLiNet [26].

Before the COVID-19 pandemic began, around March 2020, the network's members gathered for regular monthly meetings, where representatives from each member library participated to discuss and plan the network's activities. These meetings took place at different venues, adhering to a rotation system among the member libraries. Each meeting was led by the representative of the library serving as the host. Minutes were meticulously recorded during these meetings, including the allocation of responsibilities and tasks among the collaborating members. Starting from March 2019, due to the pandemic, these meetings transitioned to being conducted online, maintaining their regular schedule.

Furthermore, the network has established criteria for admitting new members, requiring that they share a common theme related to the shipping industry, which serves as the connecting factor among the existing members.

Since 2018MarLiNet has engaged in various activities:

- Introduced at the 10th QQML conference in Chania, May 2018.
- Presented at the Koventarios Municipal Library of Kozani's opening in October 2019.
- Participated in the 24th Panhellenic Conference of Academic Libraries in Larissa, November 2018.
- Featured in a workshop at the Hellenic Chamber of Shipping in December 2018, focusing on maritime knowledge alliances.
- Contributed to the IFLA conference in Athens, August 2019, through a poster session and promotional materials.
- Returned to the QQML conference in May 2020, presenting on maritime cultural heritage.
- Presented at the 26th Panhellenic Conference of Academic Libraries in Athens, December 2020, discussing the pandemic's impact on member libraries.
- Hosted a webcast on Greek shipping's role in the 1821 Revolution in April 2021.

5.3 Strategic Planning of MarLiNet

Strategic planning enables cultural organizations and collaborations to establish a robust support framework. Through creating a uniform, documented strategic plan, stakeholders across all organizational levels can make more cohesive and synchronized choices. This approach facilitates efficient resource utilization by each entity to achieve and accomplish predetermined goals [1].

The initial phase in shaping the Network's trajectory begins with crafting a clear mission statement. This statement should embody four key qualities: scope, resilience, challenge, and distinction. It needs to be comprehensive, covering all the organization's initiatives, and enduring, guiding the organization's path for at least two decades ahead [2].

On the flip side, the vision outlines an envisioned future scenario for the organization in a concise manner. It addresses two core inquiries: how the world might differ in five years if the organization persists, and the organization's part in effectuating this change.

Responses should articulate broad outcomes rather than detailing the organization's actions. An effective vision statement motivates and challenges stakeholders and team members without being too daunting [2].

MarLiNet's mission is to preserve and promote Greece's maritime cultural heritage through its collections. Its vision is to enhance services and collections by facilitating access and sharing of data with users. MarLiNet value boldness and insightfulness, cooperation, confidence, and specialization [26].

Strategic planning includes identifying shared values, which are the fundamental principles guiding the organization. While MarLiNet may not explicitly list its values, several can be inferred from its presentations, contributions, and events. These values include:

- Boldness and Insightfulness: Embracing innovation and flexibility, believing there's always a better way, and actively exploring possibilities.
- Collaboration: Sharing goals, exchanging ideas, leveraging collective knowledge and experiences, fostering honesty, and celebrating achievements as a unified team.
- Trust: Building mutual trust within the network and with the community, enabling confident decision-making and supportive partnerships.

The following is an analysis of the internal and external environment of MarLiNet using the PESTLE and SWOT analysis models (Tables 3 and 4).

Table 3. Analysis of the external environment of MarLiNet–PESTLE analysis

Category	Factors
Political	Major cuts in state budgets of all agencies, reduction in budgets of university institutions, increased bureaucracy in the case of program proposals, reduction in recruitment of qualified staff
Economic	Economic recession, more demand for services, possible changes in the economy from the impact of the pandemic on university and special libraries, lack of funding for procurement of electronic resources
Social	Impact of the pandemic on the use of network services, security and hygiene measures, ability to provide services to all communities served, lifelong learning and information center for the community served
Technological	Internet and internet databases, hosting of webinars/conferencing/community forums, provision of technology for people with disabilities, social networking and media, integrated search tools for internet searches (user-friendly, multiple databases), smart devices and gadgets (mobile phones), E-books and e-readers, teleworking and changing management, increasing online communication for organization, administration, planning and control, increasing use and demand for technological infrastructure by staff
Legal	Copyright, accessibility policy for people with disabilities, personal data, Legal status of each member organization and the restrictions imposed on its activities and services to other member libraries
Environmental	Increase in energy costs, climate change, recycling, sustainable development and green libraries

The evaluation of the objectives of the MarLiNet network focuses on increasing the availability of titles for library users, saving resources through strategic acquisitions, enhancing the readability of maritime collections, and maximizing their visibility. Following PESTLE and SWOT analyses, these goals are deemed feasible despite challenges such as budget reductions and resource limitations. Collaborative strategies, including shared acquisitions and leveraging open-source software, are suggested to address these issues. Increasing collection visibility through planned initiatives like workshops and

Table 4. Analysis of the internal environment of MarLiNet–SWOT analysis

Factor type	Description
Internal factors	
Strong points (+)	
Resources	– More resources at less cost
Collections	– Development of collections without additional costs
Relationships	– The Network builds strong relationships between member libraries
User Needs	– Meeting the maximum information needs of users
Shipping Info	– Specialized information on shipping to the right user at the right time
Events	– Experience in organizing events, seminars, workshops, presentations
Savings	– Budget and staff savings in libraries
Weak points (–)	
Electronic resources	– Increase in the cost of electronic resources
Funding	– Funding constraints
Time	– Lack of time of participating members as it is voluntary work
Organizational diversity	– Heterogeneity of the parent organizations of the organizations
External factors	
Opportunities (+)	
Consortium benefits	– Consortia always provide a better offer and more benefits
Strengths synergy	– Synthesis of the strengths of each member organization for the benefit of the readers of all stakeholders
Knowledge	– Know-how in information and communication, seamless knowledge
Technology	– New technologies
Purchasing power	– More purchasing power and synergies
E-literacy	– Evolution in e-literacy
Open data and access	– Open data and open access
Librarian role	– Changes in the role of the librarian
Work management	– New ways of working and managing
Membership	– Involvement of new members
Sponsorships	– Strong sponsorships
Threats (–)	
Member distrust	– Possible distrust among members can lead to a problem in the network's negotiations
Resource feasibility	– Purchase of bundled electronic resources is not feasible for all members
Management issues	– Management issues

seminars is also highlighted as a strength of the network, leveraging its existing expertise and international connections. Promotion efforts, supported by past presentations and events, are seen as achievable and capable of being expanded upon to further enhance the network's objectives.

This is a draft version of the proposed strategic plan of MarLiNet for the years 2022–2024 (Table 5).

Table 5. Strategic plan of MarLiNet 2022–2024

Objective	Strategy 1	Strategy 2
Expand the availability of titles to library users	Unified search service—integration of bibliographic databases, institutional repositories, and electronic sources of the individual member-libraries	Interlibrary loan—listing of lending policies, design of interlibrary loan regulation, implementation of regulation
Saving resources through a rational new acquisition strategy	Design of a single acquisition policy—scouting of international practical acquisitions, listing of members' political acquisitions, design of a single acquisition policy	N/A
Increasing the readability of the maritime collections of libraries	Promotion of MarLiNet and participation—contribution to Lifelong Learning—Information Literacy, installation and configuration of a hosting platform for IL, workshops, content development, and design of IL courses	Expanding the role of MarLiNet—discovery maritime bibliography seminars, organization of summer schools
Maximize their visibility as important collections on maritime issues	Promotion of services and collections on maritime issues MarLiNet—creation of a social media department, development of promotional material, user information seminars, approach of potential sponsorships	N/A

6 Conclusions

Synergies play a crucial role across sectors, fostering cooperation among individuals and organizations to achieve mutual success. Libraries, often part of non-profit entities, can leverage these collaborations to amplify their impact within communities. Various forms of collaboration exist, including networks, coalitions, movements, and strategic alliances. Establishing synergies involves identifying potential partners, planning, defining visions and missions, forming steering committees, setting short-term objectives, promoting consolidation, and evaluating outcomes continuously. These partnerships enable participants to enhance their effectiveness in executing actions and processes, drawing inspiration from international and domestic examples that emphasize harmonious relationships, organizational learning, and a collaborative ethos.

This paper discusses conducting a comparative analysis between a Greek and a foreign consortium of academic libraries to understand their similarities, differences, advantages, and disadvantages. The aim is to identify if there exists a gap between foreign and Greek consortia. Good practices identified through this analysis form the foundation for designing a strategic plan proposal for the case study, specifically for the Maritime Libraries Network. This initiative underscores the significance of synergies for both participating members and broader Greek partnerships and alliances. The process involves several steps, including identifying the network's current strategy, analyzing its external and internal environments, and designing a concrete strategy informed by international and national good practices. The study aims to serve as a research tool for developing strategic synergies among libraries from diverse institutions, emphasizing the importance of such collaborations in today's rapidly evolving technological landscape and global economic and social crises.

References

1. Addis, M., Rurale, A.: Managing the Cultural Business. Routledge (2021)
2. Angelica, E.: The Fieldstone Alliance Nonprofit Guide to Crafting Effective Mission and Vision Statements. Fieldstone Alliance, Minnesota (2001)
3. Blackburn, R.: Interlibrary cooperation. In: Orne, J. (ed.) Research Librarianship, Essays in Honor of Robert B. Downs, pp. 51–73 (1971)
4. Crawford, A.: The Meaning of the Library: A Cultural History. Princeton University Press (2015)
5. Faulkner, D.: International Strategic Alliances: Co-operating to Compete. McGraw-Hill (1994)
6. Greek Libraries Network: Λίγα Λόγια. https://network.nlg.gr/liga-logia/ (2020)
7. Grisham, F.: Library cooperation. J. Libr. Adm. **17**(4), 37–45 (1992)
8. HEAL-Link: About—History. https://www.heal-link.gr/en/about-history/ (2020)
9. HEAL-Link: HEAL-Link in numbers. https://www.heal-link.gr/ (2020)
10. HEAL-Link: Open Access agreements. https://www.heal-link.gr/en/open-access-agreements/ (2020)
11. HEAL-Link: Λοιπές Υπηρεσίες ΣΕΑΒ: https://www.heal-link.gr/υπηρεσίες-σεαβ (2020)
12. H.E.LI.N.: Basic principles of H.E.LI.N. http://diovi.lib.unipi.gr/index.php/en/features-mainmenu-48/2015-09-21-18-26-30 (2021)
13. HSLNKC: HSLNKC Health Sciences Library Network of Kansas City. https://hslnkc.org/ (2020)
14. Hughes, J., Weiss, J.: Simpler rules form making alliances work. Harvard Bus. Rev. (2007)
15. IAML-GR: Το Ελληνικό Παράρτημα. http://iamlgreece.teiep.gr/?page_id=545 (n.d.)
16. IARLA: About IARLA: https://iarla.org/ (2021, June 15)
17. IATUL: The History and Development of IATUL. https://www.iatul.org/about/history (2019)
18. Kaiser, M.M.: Strategic Planning in the Arts: A Practical Guide. Brandeis University Press (2019)
19. Literacy for All: About us. https://literacyforallnetwork.wordpress.com/ (n.d.)
20. M25 Consortium of Academic Libraries: About the Consortium. https://www.m25lib.ac.uk/about-the-consortium/ (2021)
21. M25 Consortium of Academic Libraries: Strategic Plan 2019–2022. https://www.m25lib.ac.uk/about-the-consortium/governance/strategy-2019-2022/ (2021)
22. M25 Consortium of Academic Libraries: Cpd25. https://www.m25lib.ac.uk/cpd25/ (2021)

23. M25 Consortium of Academic Libraries. Access25. https://www.m25lib.ac.uk/m25-resources-and-activities/access25/ (2021)
24. M25 Consortium of Academic Libraries: Benefits of Membership. https://www.m25lib.ac.uk/about-the-consortium/m25-membership/benefits-of-membership/ (2021)
25. Maritime Libraries Network: Welcome to Maritime Libraries Network—MarLiNet. https://marlinet.aegean.gr/ (2021)
26. Maritime Libraries Network: Aims—Vision. https://marlinet.aegean.gr/content/aims-vision (2021)
27. McLaughlin, T.A.: Nonprofit Mergers and Alliances, 2nd edn. Willey (2010)
28. McNamara, C.: Field Guide to Consulting and Organizational Development: A Collaborative and Systems Approach to Performance, Change and Learning. Authenticity Consulting (2006)
29. Moczulska, M., Seiler, B., Stankiewicz, J.: Coopetition in for profit and non-profit. Management **23**(2), 138–156 (2019)
30. Paquette, J., Redaelli, E.: Arts Management and Cultural Policy Research. Palgrave Macmillan (2015)
31. Wroblewski, L.: Culture Management: Strategy and Marketing Aspects. Logos Verlag (2017)

Author Index

A. Katsirikou (Ed.): QQML 2023, 68, p. 157, 2026.
https://doi.org/10.1007/978-3-032-01264-7

The manufacturer's authorised representative in the EU is Springer Nature Customer Service Centre GmbH, Europaplatz 3, 69115 Heidelberg, Germany. If you have any concerns regarding our products, please contact ProductSafety@springernature.com

Printed and bound by CPI Group (UK) Ltd, Croydon, CR0 4YY
21/09/2026
02219968-0014